The European Modernity

A part of Rubens' series on the theme of the triumph of faith or the Eucharist, which generally shows a large cross and the chalice or cup of the sacramental bread and wine shining while held aloft on a chariot of victory pulled by angels and cherubs on the left of the canvas (not shown). On the right of the canvas shown here are depicted personified the four figures who are captured and follow behind the chariot: science (man with dark hair holding armillary sphere), art (young poet's head wearing laurel wreath), philosophy (old Socratic greybeard with walking stick) and nature (nubile young woman with four breasts, marvellous but monstrous without grace) (1627).

Source: Musée des Beaux Arts, Valenciennes, France.
Painting on loan from the Louvre, Paris.

The European Modernity

Science, Truth and Method

J. P. S. UBEROI

OXFORD
UNIVERSITY PRESS

OXFORD
UNIVERSITY PRESS

YMCA Library Building, Jai Singh Road, New Delhi 110 001

Oxford University Press is a department of the University of Oxford. It furthers the University's objective of excellence in research, scholarship, and education by publishing worldwide in

Oxford New York

Athens Auckland Bangkok Bogota Buenos Aires Cape Town Chennai Dar es Salaam Delhi Florence Hong Kong Istanbul Karachi Kolkata Kuala Lumpur Madrid Melbourne Mexico City Mumbai Nairobi Paris São Paolo Shanghai Singapore Taipei Tokyo Toronto Warsaw

with associated companies in Berlin Ibadan

Oxford is a registered trade mark of Oxford University Press in the UK and in certain other countries

Published in India
By Oxford University Press, New Delhi

ISBN 019 565 5478

Typeset in Sabon
By Excellent Laser Typesetters, Pitampura, Delhi 110 034
Printed by Roopak Printer, Noida
Published by Manzar Khan, Oxford University Press
YMCA Library Building, Jai Singh Road, New Delhi 110 001

O, what a world of profit and delight,
Of power, of honor, of omnipotence,
Is promised to the studious artisan!
All things that move between the quiet poles
Shall be at my command. ...

I'll have them fly to India for gold,
Ransack the ocean for orient pearl,
And search all corners of the new-found world
For pleasant fruits and princely delicates;
I'll have them read me strange philosophy
And tell the secrets of all foreign kings;
I'll have them wall all Germany with brass.

...Stranger engines for the brunt of war
Than was the fiery keel at Antwerp bridge
I'll make my servile spirits to invent.

[If sin by custom grow not into nature, ...]
I do repent, and yet I do despair:
Hell strives with grace for conquest in my breast!

Marlowe, *Doctor Faustus*, 1588

There dwell two souls, alas, within my breast,
And each desires to sever from the other.

Goethe, *Faust I*, 1801

Preface

All, all of a piece throughout:
Thy chase had a beast in view;
Thy wars brought nothing about;
Thy lovers were all untrue.
'Tis well an old age is out,
And time to begin a new.

John Dryden, 1631–1700

Moreover, let no one be afraid that the contemplation of signs will lead us away from things. On the contrary, it will guide us to the innermost nature of things.

Leibniz, 1646–1716, *Discourse on metaphysics*

A few personal words of apology and explanation seem to be called for to present this third and final attempt, in which the results of the earlier two attempts are also included, to try and persuade some Indian social science colleagues and others not to continue to concede the monopoly of self-representation so far exercised by modern European culture and power or the European modernity. It is useless to try and explain away that monopoly as itself the expression of European power, political, military and economic domination and exploitation or imperialism, let alone simply of the late industrial capitalist addiction of consumption, although it is true that European culture is the only living one that has spread world-wide in modern times by whatever means. That simple attitude will amount only to wishful thinking: it erodes one's own sense of responsibility, and it will be ineffective as a mere ritual of exorcism since it fails to notice the essential dualism of power and culture instituted by the European enlightenment or project of modernity, which is accurately speaking a project that is better characterized as affirming a duopoly whereby modern European power and culture are indeed truly divided between themselves, the one against the other, but always reunited against us. For an everyday instance, we are regularly presented in Delhi with two separate bills for use and consumption of the medium of electricity, the one for power and illumination at 220 volts (DESU) and the other for telephonic cultural/linguistic communication at 12 volts (MTNL), and they will remain forever mutually incommensurable although one must pay for the two in a single currency of money.

On the contrary, I find that the attitude of resentment and even fervent appeals for justice against European power appear on examination to be more than compensated for by the (unfulfilled) promise of universal modern learning and universal technical culture, the enlightenment of Europe's arts and sciences, useful and useless, transplanted in Asia as Raja Ram Mohun Roy, the father of modern India, wanted them (c. 1800). We are all willing recruits, therefore, and not merely mercenaries, slaves or forced labour in the service of the Enlightenment project, classified in whichever of three grades of merit, all forever requiring European supervision. We are its appointed trained missionaries supported in the colonies and with regular metropolitan home leave from time to time, its loyal soldiers and subalterns fighting against Oriental despotism, feudalism or medieval superstition and antique obscurantism, or else thirdly, hired coolie labour or babus in its ethnic or local data collection branch. The (uncritical) missionary, the subaltern, the babu or the coolie might be of the left or of the right, liberal or Marxist, but in either case leaving aside the reactionary native traditionalist who possessed no curiosity about modernity and would not follow the leader. The other posited non-dualist method, theory and praxis of Gandhi or of the alternative Indian modernity and its counter-Enlightenment from Kabir (1440–1518) and the rise of the vernaculars onward, as I see it, are not at all occupied with the transfer of power (Rajput/Thakur) or the protection of tradition (Brahmin), but rather with the recovery and affirmation of self-rule and self-reform in culture as well as power, believing as I do that their conjoint locus is civil society, including the universities and independent institutions, and not traditions of antiquity and/or the modern Indian state, whether under the Mughals or under the British or thereafter.

In spite of some attempted self-conscious and diligent ethnological fieldwork or participant-observation abroad during 20 months of sabbatical and other leave of absence from the University of Delhi (1971, 1979–80, 1987, 1991 and 1993), I did not find my samples or soundings of the English-speaking intellectual world at all sympathetic to a project of the constitution of modern European culture or the European modernity as a coherent object of study. There appeared to be three parallel lines of argument and evidence to discourage the attempt. (a) It was politely contended, somewhat after the fashion of Max Weber, that modern European culture was an historical composite, a concatenation of institutions or complexes of circumstances, 'Christian, capitalist, democratic, constitutionalist, scientific, etc.', each of which was the subject of independent specialist inquiry and study. (b) In all of these separate studies I was incompetent, although in accordance with the principles of Enlightenment universalism I was left free to enrol as a student of any specialist branch, but always without asking if all the different branches might form a system of knowledge, i.e. in contrast with faith and action. I thought that the classification of faith, knowledge and action was itself worth study as a matter of the initial conditions or assumptions of civilization, and not to be taken as denigrated by Anglo-Saxon empiricism as only of philosophical or 'third order' interest, i.e. following behind first, the world in itself, and second, its spontaneous descriptions by the scientist. (c) There were of course already extant or under way a variety of liberal, Marxist or traditionalist (e.g. Toynbee) and post-modernist projects for the study of modern European culture, if that was indeed taken as a stable coherent object of study, and it would save both my time and ignorant labour to join and serve in one or the other of them. If I did not join any of them, I should have to beat them all, or else fall back upon so-called local knowledge, tradition or even fundamentalism as others had done before me, thereby again leaving modern European tradition, not as one among the many, but as the mirror and the measure of self-reflexivity of all the others.

I expected and received more sympathy on the continent of Europe for my supposition of a plurality of cooperating as well as competing universalisms, each of them located in but not limited by space and time (and of which European critical rationality might well be celebrated as one). I shall try to make the most of it in the pages that follow, although one cannot honestly say that repeated ethnographical visits made during the aforesaid years outside the English-speaking world to Paris, divided and reunited Germany, Zürich and Warsaw (and also Tokyo, 1999) strengthened my resolve. I did learn some post-modern and other critiques of modern European culture and power, but on the other hand began to despair of the search for self-renewable alternatives (e.g. Foucault), without which we face the indefinite prospects of democracy without ideology, religion without magic, theology or institutions, science without ethics, power without purpose and modernity without end. There seemed to be nothing for it but to try and find one's own way into and out of the modernity component of (say) modern industrial capitalism, the other parts of which civilization were already sufficiently well known to us.

I owe a sincere apology, therefore, for now attempting to discover or invent, beyond the limits of my specialist incompetence and self-evident lack of means, not only one but *two* Europes for study, which had been previously presented separately, the official and orthodox modern dualist one of the Enlightenment and the other Europe, here identified as the heterodox or occulted Hermetic tradition and the radical European underground, of course invisible to the modernist regime except as anarchist, reactionary, heretic or treasonable just like the original Gandhi. My conclusion will be that the duopoly or the dualism of the European modernity, i.e. of fact and value, truth and reality, theory and praxis, howsoever efficient in calculation or enlightened in intention, is altogether bad for humanity because a person whose head, heart or hand (so to say) habitually acts without first consulting the other two organs is very likely sooner or later to produce something misshapen and horrible, either an Auschwitz or a Hiroshima or both, whether this process of faith, knowledge and action appears as expediency (the end justifies the means) or alternatively as technicism (the means justifies the end).

The explanation to be offered is that I have adopted the non-dualist point of view and attitude in relation to method, theory and data in this book of the scientific non-specialist and the non-expert/non-elite who, rejecting the two earlier hypotheses (or excuses) of incoherence of the object and subordination of the subject, will try and regard the project of the European modernity, not in the given terms of (a) an historical concatenation of circumstances or an evolutionary chain of causes, or (b) another type of society with its complexes of distinctive features, or (c) a mechanical, organic or semiotic system of causes, factors and effects, structures, processes and functions or even of ends and means, but simply as (a) a universe of discourse, explaining the language or the *medium*, *code* and *message* of the modernity project, and attempt (b) to discover and define its *conditions of existence*, what are called in philosophy its presuppositions and assumptions or in science its initial situation and boundary conditions and in politics the limits of its art of the possible, so as (c) to better understand their role, meaning and effect as what are in mathematical science termed 'constants'. The chief historical turning points here supposed begin firstly with the coming together in western Europe of the elite Renaissance of the south and the popular Reformation of the north after c. 1500, conjointly progressing to the scientific revolution and the political Commonwealth in England (c. 1650), then flowering simultaneously in the French revolution and the Industrial Revolution in England (1760–1840), the expansion of imperialism as a system of power and the diffusion of the Enlightenment as a pattern of culture throughout the world, and finally ending simultaneously with the implosion of Auschwitz and the explosion of Hiroshima, which I shall argue formed one civilizational event (1945). We will not attempt an historical exposition of the European modernity, however, but simply suppose all these several turning points to coexist in simultaneity rather than in succession as far as possible. The discourse of the Enlightenment, for example, would be implicitly assumed to be pegged down to the trinity of, say, Locke in England, Voltaire in France and Kant in Germany, the first of whom was born in 1632 and the last died in 1804, and who form a modern set of method and theory without any further word from me.

Chapters 2 and 4 following thus attempt to explain, using the monographic method of exposition, the origin and the end of the European modernity in c. 1500 and 1945, respectively, from the new Eucharist to the bomb, with a superadded section on the meaning and effects of time and technology during the Cold War. Modern European culture, like any other, especially after the rise of the state in prehistory, has always a bimodal distribution of existence, the one manifest in peace and the other in war. My defence for trying to define both modes of existence under a single semiological theory, at a time of the changing of gear from the one state to the other, is only that it is better for social science to have so tried and failed than not to have tried at all. The rest of the book, consisting of chapters 1, 3 and 5, is devoted to exploring what I call the other Europe and its alternative tradition of modernity within, the other self of the European modernity, using the same method of exposition in relation to examples chosen from the Hermetic cosmology and the sciences of nature, Dante to Goethe, the unity in variety of pluralism and civil society in India. The successive five chapters of this book thus try to discover simultaneously both (a) the official modern relations of God, man and nature or rather the modern Enlightenment's failure to find any unity in this trinity of terms, and (b) the outline of some possible alternatives from within the other Europe, pre-modern or non-modern, and from some others of Europe, namely the savage, the child and India.

In the spring of 1991 I had delivered in succession to the renowned Ivan Illich, who was the first and whose shoes I could not fill, the Karl Jaspers lectures on the theme of 'The other Europe and its arts and sciences for the world' at the University of Oldenburg, Germany, at the invitation of Rudolf prinz zur Lippe, professor of philosophy, who has been unfailingly kind, loving, efficient and gracious for ten years to help me find utterance, and who gave me his companionship as well as a long rope to rewrite the lectures as and when opportunity offered between other calls on my time. The seminar that I ran alongside the lectures during that summer semester would not have been possible without the generous support of Barbara Reiners, Rüdiger Schmidt and Karen Jürgens-Byram, my translator.

I have to thank for official as well as personal help specially the Foundation of Lower Saxony as well as the University of Oldenburg; Anke Wiegand-Kanzaki, director, and Beate-Ursula Endriss of the Haus der Kulturen der Welt, Berlin; Hartmut Rudolph, formerly with the Paracelsus edition, Marburg, and now with the Leibniz edition (Potsdam) of the Academy of Sciences, Berlin; Sabine Solf, academic director, Herzog August Bibliothek (Renaissance), Wolfenbüttel; and for his friendly post-modern instruction without didacticism Dietmar Kamper, professor of sociology at the Free University, Berlin. I am very grateful to Clemens Heller and Marc Augé for the opportunities afforded me in 1980 and 1987 to study the Hermetic tradition of knowledge respectively as visiting professor, Maison des Sciences de l'Homme, Paris, and as directeur d'études, École des Hautes Études en Sciences Sociales, Paris; and also specially to Andrew Wiercinski, Christopher Byrski and Arthur Karp, not to forget Khalid Tyabji the unforgettable, for happy times of learning in first an unhappy and then a happy University of Warsaw.

It is necessary to say that, out of the six Karl Jaspers lectures given, one has been dropped here as inappropriate to my theme of the European sciences and arts in the period of modernity. It was on the subject of Antigone, the sister and daughter of Oedipus in the play of Sophocles' trilogy, and the problem of femininity, a favourite theme of mine, and God willing it may be published separately elsewhere in good time!

Meanwhile, I remember with gratitude the unending discussions in Delhi on the question of modernity, *pro* and *con*, with André Béteille, Bhupinder Singh, Veena Das, Dharampalji, Ramu Gandhi, T. N. Madan, Vivek Minocha, Ashis Nandy, Rabindra Ray, A. K. Saran, Richard Saumarez Smith, Romila Thapar, Janet Downs Tournière, Srinivas Veeravalli, Nirmal Verma, Shiv Visvanathan, Punam Zutshi and of course everyone at the first congress of traditional sciences and technologies of India (Bombay, 1993). It is entirely owing to the intervention in 1997 of Mrinal Miri, professor of philosophy and director,

Indian Institute of Advanced Study, Shimla, where I had visited to give three lectures on the topic, that these pages appear in print after the point where commercial publishers saw no market for (independent) Indian writing on the European modernity and its other self.

Delhi School of Economics
September 1999

Contents

Plates

Figures

Figures

Plate 1: Dante is guided by Virgil through the centre of the world, the dark and the devil. The two of them undergo several rotations of symmetry before emerging again from this half of the cosmos (evil).

Source: Dante's divine comedy, fifteenth century illuminated manuscript, Venice, Biblioteca Nazionale Marciana, *Cod. It.* ix, 276 (=6902). *Productions Liber SA and Éditions Minerva SA, Fribourg-Genève, 1979*

Plate 2: Beatrice shows Dante the cosmos and answers his queries so that he understands how its good half is centred on light, the heavens and God (detail).
Source: Dante's divine comedy, fifteenth century illuminated manuscript, Venice, Biblioteca Nazionale Marciana, *Cod. It.* ix, 276 (=6902).
Productions Liber SA and Éditions Minerva SA, Fribourg-Genève, 1979

1

Symmetry and the Hermetic Tradition

That allegorical description of Hermes pleaseth me beyond all the Metaphysical definitions of Divines.... Thus there are two Books from whence I collect my Divinity; besides that written one of GOD, another of His servant Nature, that universal and publick Manuscript, that lies expans'd unto the Eyes of all.... Nature hath made one World, and Art another. In brief, all things are artificial; for Nature is the Art of GOD.

Sir Thomas Browne, 1605–82, *Religio medici*, 1: 10, 16

For the rest, the art of combinations in particular is, in my opinion, the science which treats of the forms of things or of formulae in general (it could also be called generally the science of symbols, or of forms, *characteristica sive speciosa*).

Leibniz, *Of universal synthesis and analysis*, c. 1683

I

THE VISION OF DANTE

We may begin this investigation, semiological as well as experimental, of the other Europe with a consideration of the principle of symmetry, chosen by way of example, which is to be found both in the manifest dualist forms of Europe and in its non-dualist form in the underlying, occult or forgotten and unacknowledged Hermetic tradition. The former manifest uses of symmetry rely upon the twin principles of separation *and* difference combined while in the latter tradition the emphasis is on the principles of correspondence and/or complementarity, which should not be reduced by us to merely homogeneity and/or heterogeneity.

The three component parts or aspects of symmetry in its simplified sense are the processes or operations of (a) reflection, (b) rotation, and (c) repetition, iteration or translation, etc. I think that the principle of reflection is perhaps more important, widespread and useful in nature as well as culture than is generally supposed (e.g. Weyl 1952). One may refer to the complementarity of left and right, for example, the principle of handedness, which is found even in chemical molecules, mirror reflection or what is called bilateral symmetry in animal biology. Indeed the whole language of symmetry can be regarded as simply the generalization of the laws and phenomena of mirror reflection. Similarly what we call the principle of correspondence, e.g. in the relation of the macrocosm and the microcosm, is very well known, celebrated and embodied in traditions of cosmology all over the world. It is much too often interpreted and explained one-sidedly, however, to rest upon the principles of analogy, similarities, parallelism and resemblances alone, in effect to the virtual exclusion of differences or principles of exchange, homology and what I call complementarity. The partiality shown to *relation by similitude* over, above and against *relation by difference* is characteristic of modern European scientists and scholars including, for example, the archetypal psychologist Jung, who does so apparently on the authority of the mystical or the Hermetic formula applied to cosmology, 'As above, so below.' I will try and reformulate the problem of reflection by supplying evidence of it as a very general trans-systems phenomenon suitable for interdisciplinary study in terms of its two polar aspects of similarity versus difference, and its two polar aspects of separateness or distantiation versus proximity or contiguity.

No one will be surprised and some might be even delighted, I hope, if the selected evidence itself with very little commentary from me leads to the conclusion that the eye of the beholder, i.e. the subject as well as the object, is prerequisite to the moment of reflection as a principle of symmetry or of correspondence and complementarity. This is a way of illustrating that the processes of human perception, cognition and volition form three parts or steps or phases of a single unified or dialectical cycle of self, the world and the other. Alternatively, one may say in classical language that if the process or the operation of self-rotation is characteristic of the celestial sphere (macrocosm), and that of repetition or translation is characteristic of the terrestrial sphere (microcosm), then the *moment of reflection* surely holds the key to the relation between the two spheres, if there is any.

The vision of Dante, which he started to write down in AD 1302 it is said as 'one who had lost the world and gained his own soul', is the best single representation of the world-view of the Christian middle ages of Europe or certainly it is one of the best. His great epic poem *The divine comedy* is divided into three exactly equal parts, each division consisting of 33 cantos, apart from the first opening canto of invocation which is the entry into the narrative. This makes a total of 100 cantos devoted to Dante's three journeys respectively into hell, which is of course traditional, and purgatory, which Dante had newly invented, both journeys in the company of the Roman poet Virgil, and then last into paradise, whither he is conducted by his beloved the pious Beatrice.

Dante's first journey into hell's infernal region ends in its last canto and the transition into purgatory begins with what I call the *moment of reflection*, which is here effected in the sense of a kind of inversion of the subject. Dante has now turned completely around in his orientation to the world, we are told, which is 'symbolical of his conversion from sin'. Dante and his guide had previously descended by many and varied stages to the very bottom and centre of the terrestrial world, one of the two contrasted poles of the cosmos as a whole. 'And in the midst is Lucifer, at whose back Dante and Virgil ascend, till by a secret path they reach the surface of the other hemisphere of the earth, and once more obtain sight of the stars.' This first moment of reflection or inversion in the epic, as I see it, occurs as we pass through the point that is simultaneously the centre of the earth, the world and of the devil himself, the point around which the rule of symmetry is to be preserved (see plate 1). Therefore, it describes the operation of rotation in other words through 180° or a half-circle around that central point by the participant-observer Dante who is following his guide Virgil (Hell 34).

> 'Lo!' he exclaim'd, 'lo Dis [Lucifer]; and lo the place.'...
>
> Soon as he reach'd the point, whereat the thigh
> Upon the swelling of the haunches turns,
> My leader there, with pain and struggling hard,
> Turn'd round his head where his feet stood before,
> And grappled at the fell as one who mounts;
> That into Hell methought we turn'd again.
>
> ...I raised mine eyes,
> Believing that I Lucifer should see
> Where he was lately left, but saw him now
> With legs held upward. Let the grosser sort,
> Who see not what the point was I had past [the groin],
> Bethink them if sore toil oppress'd me then.

Dante then asks his guide out of curiosity for an explanation of what has happened in that moment to his orientation in space and time:

> Where is now the ice?
> How standeth he in posture thus reversed?
> And how from eve to morn in space so brief
> Hath the sun made his transit?

To this Virgil replies:

> Thou deemest thou art still
> On the other side the centre, where I grasp'd
> The abhorred worm that boreth through the world.
> Thou wast on the other side, so long as I
> Descended; when I turn'd, thou didst o'erpass
> That point, to which from every part is dragg'd
> All heavy substance. Thou art now arrived
> Under the hemisphere opposed to that,
> Which the great continent doth overspread
> And underneath whose canopy expired [Jesus]
> The Man, that was born sinless and so lived.
>
> (Dante 1908: x, 144ff)

Similarly the second moment of reflection or inversion occurs near the end of the third part of *The divine comedy*, the ascent into paradise, where Dante's guide is Beatrice and he has parted from Virgil (see plate 2). It also appropriately occurs when we arrive at the opposite pole of the cosmos, the celestial as against the infernal, the two contrasted poles that form the single whole chain of creation, the view now having moved from the devil to God himself at the centre. Finally the movement of rotation is now entirely that of the *object* as against that of the *subject*, with whose point of view the epic had begun at sunrise on Good Friday in AD 1302 'in the midway of this our mortal life'. The subject and the object therefore again form together a kind of symmetry by themselves as the two poles of a single creation or cosmology, and also in relation to the structure of the epic poem in space and time as a narrative.

We are now in the farthest or the ninth sphere of heaven, the *primum mobile* or the first principle of motion, which is the starting point of nature and the opposite end of the world from that of the fallen angel Lucifer. The principles of correspondence and complementarity to be shown and explained to Dante are to the following effects. (a) That the first, the smallest and the innermost angelic circle, the Seraphim, answers to or controls the movement of the greatest and the outermost ninth sphere, the crystalline, outside the sphere of the fixed stars. The centre of the system of angels is of course God, while the centre of the system of spheres is the earth or Lucifer, and the two systems are inter-linked like a single system of gears in motion—but with one vital difference. Dante's new guide, the pious Beatrice, has to patiently explain to him one difficulty in the principles of movement of what the physicist would call the fluid vortex ($V_\theta = C/r$) versus the solid disc ($V_\theta = \omega r$), forming a reciprocal symmetry of velocity. (b) That the angelic circles centring on God, from the first to the ninth, are swifter and brighter the nearer they are to him, whereas their respective corresponding wheeling spheres centring on the earth, from the ninth to the first, are conversely or inversely 'more divine the farther they are from their centre' (Paradise 28).

> A point I saw, that darted light....
> There wheel'd about the point a circle of fire,
> More rapid than the motion which surrounds,

Speediest, the world. Another this enring'd;
And that a third; the third a fourth, and that
A fifth encompass'd; which a sixth next bound;
And over this, a seventh, following,...

Beyond the seventh,
Ensued yet other two. And every one,
As more in number distant from the first,
Was tardier in motion: and that glow'd
With flame most pure, that to the sparkle of truth,
Was nearest; as partaking most, methinks,
Of its reality.

This is Dante's cosmological vision of the divine essences that he was permitted to behold but he still decides to put his one difficulty to Beatrice.

It were enough; nor should I further seek,
Had I but witness'd order, in the world
Appointed, such as in these wheels is seen.
But in the sensible world such difference is,
That in each round shows more divinity,
As each is wider from the centre [of the earth].

...I must know
Wherefore such disagreement is between
The exemplar and its copy.

Beatrice gives Dante her explanation in Platonic terms that the sensible world is surely a copy or rather an image, but suitably inverted in its order and motion, of the intelligible world. She confirms that, on the one hand,

...Every orb [of nature],
Corporeal, doth proportion its extent
Unto the virtue through its parts diffused...
Therefore the [ninth] circle, whose swift course enwheels
The universal frame, answers to that [first]
Which is supreme in knowledge and in love.

The correspondence or the complementarity of order and motion between the two halves of the cosmos is therefore arranged according to the series, the intelligible world: the sensible world as 1 (the Seraphim): 9 (the *primum mobile*); 2 (the Cherubim): 8 (sphere of fixed stars); 3 (the Thrones): 7 (the planet Saturn); etc., the numerals of each pair adding up to the number ten, which is the first completion of the cycle of Arabic-Sanskrit numerals. Something does remain strictly invariant in the cosmos, however, through all such transformations from the intelligible to the sensible, the spiritual to the corporeal, and that is the rule of virtue, proportion and symmetry, qualitative as well as quantitative, as Beatrice concludes pointing toward the divine archetypes, which are essences or possibilities versus existences.

> Thus by the virtue, not the seeming breadth
> Of substance, measuring, thou shalt see the Heavens,
> Each to the intelligence that ruleth it,
> Greater to more, and smaller unto less,
> Suited in strict and wondrous harmony.
>
> (Dante 1908: 417ff)

II

THE HERMETIC TRADITION

The Hermetic tradition, taken as it is understood by specialists today, refers to that body of religious, philosophical and cosmological teaching which was reduced to writing first around AD 300 and assumed its present form as a compilation in the fourteenth century, becoming widely known notably through the publication of a Latin translation by Ficino (1471). This extant corpus of texts in Greek and Latin, the *Hermetica*, is ascribed to the non-dualist mystical theosophy of Hermes Trismegistus (the 'thrice-greatest' Hermes). (a) Above all else it teaches and celebrates unity in variety in the relations of God, man and the world, the single 'chain of being' or fabric of existence by which the totality of creation obeys the creator. (b) It comprises essentially a composite blend of the syncretic Neoplatonism of Alexandria with early opinions on what is now called natural science and its underlying principles, all written down in the critical period of the struggle between 'paganism' and Christianity. The original Hermes was one of the twelve Olympian gods, a young son of Zeus and Maia, and was represented by the Greeks as the herald or messenger of the gods and as their interpreter (*hermeneutes*), a semiotic intermediary between the celestial world and the terrestrial world who also conducted the souls of men upon death into the underworld. Hermes was further identified by the Greeks with the earlier Egyptian Thoth, the god who represented wisdom and science, and the followers of the Hermetic tradition supposed that Pythagoras and Plato got their science and philosophy from the priests of Egypt, who got them from the books of Thoth, the divinely inspired originals of the *Hermetica*.

For the Hermetist it is man's divinely allotted function to contemplate (and to continue or to complete on earth) the works of God, 'for it is in God that nature has her being' (Hermetica 1985: I, 149). As a result Hermetism as a system always properly finds a twofold affinity: (a) with the pursuit of *gnosis*, the kind of knowledge that involves union of the subject with God on the one hand; and (b) with the cultivation of experimental and scientific interests (even if they are dismissed nowadays as occult arts and sciences) on the other hand, e.g. 'spagyric' medicine (Paracelsus) as against purely spiritual or 'theurgic' medicine. In modern times even its avowed devotees like René Guénon in France complain that the Hermetic tradition is concerned with the kind of knowledge that is 'not metaphysical but only cosmological', i.e. not with the spiritual sciences which belong properly to Christ but with other and lesser 'intermediary' or hybrid sciences. I should be inclined to argue on the contrary that it is precisely this non-dualist character of Hermetism—mystical, theosophical, symbolic and mathematical, experimental and scientific—that can be a source of strength at the present conjuncture of history and society. At any event one may note that Hermes was further identified in the Islamic tradition with the antediluvian prophet Idris or Uhnukh/Enoch, who some say is the same as the Buddha or more accurately Budha (Wednesday, the planet Mercury) in India, and so forth.

What is the relation, or is there any, between the Hermetic tradition and the cosmology of Dante? I shall present five or six passages selected from the *Hermetica* that support the existence of a relation specially concerning two topics: (a) the unity of opposites as a principle of mind and the world; and/or (b) the moment of reflection encountered in passing from the one to the other of any pair of terms.

The first text adduced here is on the *structure of contrariety* in the world and is taken from the Corpus Hermeticum book 2, a discourse of Hermes Trismegistus to Asclepius (the son of Hermes) in Greek dealing with some difficulties about the nature of space (*topos*).

Ascl. How is it then, Trismegistus, that the things which in our world move other things are moved together with the things they move? For I have heard you say that the planet-spheres are moved by the sphere of the fixed stars; (and surely that sphere is itself moved). —*Herm.* In that instance, Asclepius, the two things are not moved together. Their movements are contrary; for the sphere of the fixed stars is not moved in the same way as the planet-spheres, but in the opposite direction. And the contrariety of the two movements keeps the fulcrum stationary; for motion is stayed by resistance. The planet-spheres then, being moved in the opposite direction to the sphere of the fixed stars.... It cannot be otherwise. Look at the Great Bear and the Little Bear. As you see, they neither set nor rise; are they moved, think you, or do they stand fast? —*Ascl.* They are moved, Trismegistus. —*Herm.* And of what kind is their movement? —*Ascl.* It is a movement which circles round one point. —*Herm.* Yes, and their revolution round one point is a movement that is held fast by immobility. For revolution round one point prevents departure from the orbit; and the prevention of departure from the orbit results in revolution round one point. And even so it is that movement in contrary directions is steadfast and stable, being kept stationary by the contrariety. I will give you an example which you can see with your own eyes.... All movement then takes place *within* something that stands fast, and is caused *by* something that stands fast (Hermetica 1985: I, 137f).

It is only fair to say that this Greek passage is difficult to decipher and Scott, the editor and English translator, marks it therefore as a later uncomprehending interpolation, but his successor Ferguson I think succeeds in making sense of it. The argument is in the tradition of Platonist-Stoic physics and it is intended to prove that space (and not the void) encircles as well as moves the cosmos, which by virtue of its self-rotation or circular orbital motion does not change place. The stars do have continuous opposite circular motions or courses like the opposite movements of planets and stars but these motions cancel each other out, i.e. they have no local (translatory) motion so that the whole cosmos remains in and for the Hermetist is moved by a stationary Topos, 'space the container' identified with 'space the mover', something rather like Plato's world-soul in the dialogue of *Timaeus* and making space like the second god Aeon. In other words, a rectilinear motion necessarily entails change of position (what is called translation) but not so the circular motion which is self-rotation in the same place, since the going round as it were cancels the going beyond. You can see the great and little Bears circling in the sky; that is rotation or circular motion, and the circular motion is kept in place by immobility or what I call self-rotation. The singularity of this argument according to the modern understanding lies in the stress laid upon *opposite* circular motions; and this stress on contrariety and oppositeness had of course already an ancient history from Heraclitus through the Stoics to Philo the Jew (Ferguson *in* Hermetica 1985: IV, 360ff).

Secondly, the principles of the unity as well as the struggle of opposites are found elsewhere stated explicitly in the Corpus Hermeticum book 10 and book 11.

But knowledge (is the perfection of science, and science is a gift of God; for all science) is incorporeal; the organ which it uses is the mind itself; and the mind is contrary to the body. A soul then, when it has entered into a body, admits into itself both things of the mind and things material. It cannot be otherwise; for all things must needs be composed of opposites and contraries.

(The things that are immortal, like the Kosmos, have in them more soul than body; but the things that are mortal have in them more body than soul.)

And all things are filled with light;...for by the friendship of contraries, and the blending of things unlike, the fire of heaven has been changed into light, which is shed on all below by the working of the Sun; and the Sun is the begetter of all good, the ruler of all ordered movement, and governor of the seven worlds.... The movements are diverse and many, and the bodies differ one from another, but there is one ordered system which extends through all; therefore, there cannot be two or more makers.

These same laws or principles that obtain in the world or in the object are also good for the subject (self). Mind (*nous*) evidently instructs our Hermes directly that you must 'bring together in yourself all opposites' in order to apprehend God. Man consists of opposites in combination, and this is a law that applies to all things alike (Hermetica 1985: I, 193f, 213, 221; II, 247, 307f).

Thirdly, in the Latin Asclepius, which is the second part of the extant *Hermetica*, it is stated that 'there are two images of God; the Kosmos is one, and man is another' (just as the macrocosm and the microcosm are but two correlated aspects of the divine archetypes). Man is the one appointed to fulfil the 'tasks of service and of worship' for the sake of which duties he is of a dual nature.

For you must note that man, in order that he may be fully equipped on both sides, has been so fashioned that each of his two parts is made up of four elements; and so, in respect of the divine part of him, which is composed of other and higher 'elements', so to speak, namely, mind, intellect, spirit, and reason, he is found capable of rising to heaven; but in respect of his material part, which consists of fire, water, earth, and air, he is mortal, and remains on earth, that he may not leave forsaken and abandoned all things that are entrusted to his keeping.

[Alternatively] Man has a pair of hands and a pair of feet, as bodily members, that he may therewith do service to the lower and earthly part of the universe; and he has four mental parts, namely, reason, intellect, memory, and foresight, that he may therewith know and reverence all things divine (Hermetica 1985: I, 305f, 307n).

The editor Scott agrees that taken either way this double set of elements, between the two aspects of which there is to be established a relation of analogy or homology as we shall see in fig. 1 and which occurs in several places, is characteristic of Hermetism. Ferguson adds importantly that it is not simply a bipartite division, however, between the lower and the higher in the manifest physical universe, the sublunary versus the celestial; rather it is something more abstract like determinism versus free will or the operation of grace. It is a contrast between (a) the realm that is subject to the rule of Heimarmene or natural law working through the seven planetary spheres identifiable with the god Aeon, and (b) the free realm that is beyond the rule of Heimarmene, i.e. beyond Fate, Fortune or Destiny and the influence of the heavenly bodies (Hermetica 1985: II, 33; III, 56, 108f; IV, 357, 386, 402f, 430, 484).

I may interpose here that the kind of symmetry which we have noted above of the nature of space also holds good of the world's time for 'time in its ordered course renews all things in the Kosmos by alternation' (and not by linear progression). This refers specifically to the regular succession and rhythm of the seasons in the microcosm (the annual cycle, terrestrial); and to the eternal return of the stars to their appointed places (the great Platonic year, celestial), so that the stars are always both before and after themselves in their forms of rhythm and symmetry and are so bound into a unity by orderly cycles of time (Hermetica 1985: I, 351; III, 65; IV, 420, 430).

Fourthly, there is the Hermetic belief that in God the two procreative functions of the male and the female are eternally united—'God is bisexual' for the Hermetist. In mortal races of course the two sexes are separated and divided, each individual being either male or female; but in the act of procreation, which is both a mystery and a sacrament, this separation ceases and male and female become one, as God is one. This happens, I may explain, not only or chiefly through their achieving homogeneity or uniformity or by the division of labour and the unity of its product, but rather by the mutual exchange of roles or even of selves.

For if you note that supreme moment when, through [mutual]interaction without pause, we come at last to this, that either sex [which is filled with procreative force] infuses itself into the other, the one giving forth its issue, and the other eagerly taking hold on it and laying it up within, you will find that at that moment, through the intermingling of the two natures [in this sacramental act], the female acquires masculine vigour, and the male is relaxed in feminine lànguor (Hermetica 1985: I, 333f; III, 143).

Fifthly, we may conclude this short survey of the moment of reflection in the *Hermetica* with an excerpt from Kore Kosmu, the 'pupil of the eye of the universe', another holy book of Hermes Trismegistus in Greek. It assumes what is a topologically inverse relation between the earth and heaven or the sky, the feminine and the masculine, but without upsetting the symmetry or sympathy that makes the climates of earth answer to the climates of heaven—and it ends with a nice Egyptian touch, although the earth was feminine in Greece rather than in Egypt.

The Earth, said Isis [to her son Horus], lies in the middle of the universe, stretched on her back, as a human being might lie, facing toward heaven. She is parted out into as many different members as a man; and her head lies toward the South of the universe, her right shoulder toward the East and her left shoulder toward the West; her feet lie beneath the Great Bear, and her thighs are situated in the regions which follow next to the South of the Bear.... But the right holy land of our ancestors [Egypt] lies in the middle of the earth; ...and that, my son [Horus], is the reason why the men of this land, while they have in equal measure all other things that all the rest possess, have this advantage over all other men, that they are more intelligent. It could not be otherwise, seeing that they are born and bred upon Earth's heart (Hermetica 1985: I, 501f; III, 570; IV, 467).

In the classical Greek versions of Hesiod, as indeed in all Indo-European mythology, the original *separation* of the sky and the earth is somehow required in order to inaugurate the cosmos (a) that is ruled by pairs of opposites which conflict as well as harmonize with each other, i.e. the laws of correspondence, contradiction and complementarity; and (b) where all living beings are born, take shape and survive through the separation and the union of opposites, competition and cooperation. In the beginning, even before the time of the Olympian gods, there was the 'deep-breasted' earth, Gaia the omnipotent, from whom all things issued forth opposed to chaos. But initially the starry sky Uranus, her first-born son and husband, still covered her entirely so that their offspring could not develop in the light of day 'for want of distance between their two cosmic parents'. Gaia then became angry and persuaded her last-born son Kronos to castrate his father Uranus, so that forever afterwards it is only but once a year that the sky will fecundate the earth, and each of them will otherwise remain in its separate assigned place (Vernant 1972: 85f).

The unity and the conflict of opposites, in other words, necessarily require some degree of separateness or distantiation between the pair of two opposed terms if their relation is not to collapse into the original confused togetherness and immobility of primeval chaos, which was the condition of the cosmos before the emergence of structure via the mother Gaia. At the very beginning of the act of creation, everything flew apart into separateness *and* difference, just as conversely everything will collapse eventually into homogeneity (togetherness in the sense of proximity *and* similarity) in the final moment of the apocalypse. But in the period of the present, life and the cycles of time, terrestrial and celestial, answer to the order of the cosmos through 'structure', i.e. all that has form including metamorphosis or formation, transformation and malformation, which emerged in the space duly opened between heaven and the earth, alternately revealed and concealed by the rhythm of the day and the night.

III

PARACELSUS

> . . . Man is all symmetry,
> Full of proportions, one limb to another,
> And all to all the world besides.
> Each part may call the farthest brother,
> For head with foot hath private amity,
> And both with moons and tides.
>
> George Herbert, 1593–1633

It will be evident that this whole single unified cosmological chain or fabric of self, the world and the other, as specially emphasized anew by Paracelsus and others in updating the Hermetic tradition, was internally subdivided into two incomplete halves, the macrocosm and the microcosm or the warp and the weft, between which the relation to be sought is that of *homology* (e.g. polarity) rather than that of *analogy* (parallelism) as modern European commentators favour it. The unity in variety of nature and the unity and continuity of the great chain or fabric of being ultimately depend, not only upon the principles of similarity, homogeneity, resemblance, correspondence and sympathy, but also equally upon the contrary principles of difference, separateness, heterogeneity and antipathy or complementarity, the two aspects working together by the Hermetic 'friendship of contraries' and its 'blending of things unlike'.

We may say in very general terms that this macrocosm/microcosm or warp/weft relation is in the tradition of Plato and Pythagoras versus the official tradition of Aristotle and the Peripatetic school, which always emphasized on the contrary the discontinuity, in the sense of separateness *and* difference, between the two halves of the cosmos, e.g. that the circular motion (rotation) is natural to the celestial world, the unchanging heavens, while rectilinear motion (translation) is natural to objects of the changing sublunary world, the world of generation and corruption, etc. The principles of symmetry or of the Hermetic correspondence and complementarity are then special aspects of the archetypal unity and struggle of opposites; and the forms, rhythms and symmetries of the macrocosm and the microcosm stand in or express the relations of their equality, polarity and progression in the general sense of dialectics. What I have termed the *structure of homology* between the two halves or aspects is together constituted (a) by their separateness, division and distantiation as well as by proximity and contiguity, (b) by their mutual differences as well as by similitude or correlation, and (c) by mutual opposition or antagonism as well as by mediation, affinity and complementarity or exchange, for the golden rule common to continuity and discontinuity in the interwoven fabric of warp and weft is that 'there is one ordered system which extends through all'.

This Hermetic dialectics of the macrocosm and the microcosm or the warp and the weft is rather like the age-old symmetry of the two lovers celebrated in the latest popular hit, 'Upside down', where Diana Ross complains in simple paraphrase, 'Boy, you turned me upside down, inside out and round and round' (1980). I adduce this example of song because the union of opposites in this sense is expected to be happy and fertile in the unity of its product, the interwoven fabric, whereas the union of similars is found to be after all only temporary, infertile or unproductive. This result is of course very unfortunate for those people, and there appear to be very many, who would like to define equality as sameness, i.e. as necessarily entailing togetherness as well as similitude, but Diana Ross and I are not to blame for it: we had voted for combining equality with difference.

The Hermetic tradition, whatever its underground life as paganism or Christian heresy during the middle ages of Europe, was developed in the early modern period with a new profound scope, vision and practice by the life and work of Paracelsus (1493–1541) in the sciences of medicine, physiology, chemistry, etc., as well as in religion and philosophy, not to say also in relation to society and politics. Paracelsus, the Swiss-born wandering Christian doctor who among other things stood firmly with the national peasantry during their uprising (1525) and again with the Anabaptists under persecution and the Black Death in the Tyrol, should be regarded on the grounds of his wide and deep influence as having played the role of the midwife of an alternative non-dualist modernity in relation to the new aspects of faith, knowledge and action or of science, religion and politics which combined the Renaissance and the radical Reformation north of the Alps. Even if the great project failed in relation to power, became recessive in science or occult in culture and went underground again for the future, its outline and contours are visible, for example, during the period of the Commonwealth in England, 1649–60. The historical sources and antecedents of Paracelsus for such a method of approach in philosophy as well as science, which he applied perhaps for the first time to the knowledge of nature in a thorough empirical way, have been discussed and reviewed by Pagel (1958) in detail. They are of course Gnosticism and the Cabala, alchemy, Neoplatonism, the popular Puritan pantheism of the Beghards and the Beguines or the Brothers of the Free Spirit, Salomo ibn Gebirol (the Latin Avicebron), Lull, Nicolaus Cusanus, Agrippa of Nettesheim. All of them have something of the non-official occult and the European underground, regardless of their precise dates, and may be said to be all connected with the Hermetic tradition in its radical aspect, if a single label is required for the sake of convenience and which should combine certain aspects of the Renaissance as well as the Reformation. It is this latter combination of course that makes Paracelsus so much more than a follower of the *philosophia adepta* (philosophy of the initiate), which he had learnt from his father, and into the radical reformer, 'physician and apostle of the new creation' (Rudolph).

Paracelsus was perhaps the most original of the radical Reformers, as they are now known, in his attempt to update Plato, so to say, and to reunite the study of nature, specially of medicine and the arts of healing, with cosmology as well as chemistry and anthropology. With his truly heroic original vision, Paracelsus opposed the established traditional explanation of phenomena in terms of the four elements, humours and qualities of the sublunary world as distinct from the celestial world (Aristotle). He proposed to put in their place the specific properties of individuals and species, whether of creatures, objects or events, e.g. diseases, and he rejoined the terrestrial world with the celestial world, self with the other, through the Hermetic cosmology. There are as many classes of diseases, he said, as there are organs of the body for the 'worms of the marrow are different from the worms of the gut'; and the disease is a specific event taking place in the organ. A corresponding planet in the heavens will then in this Hermetic language indicate to the physician the specific metal from which to prepare his medicine (iatrochemistry). Thus according to Paracelsus (a) anaemia among the diseases should be treated with (e) iron among medicines *because* (b) blood is among the organs as (c) Mars is among the planets which is as (d) iron is among the metals. This mode of thought, knowledge and action will take us in the service of healing the other from the human body up to the stars in the heavens and then back again to the terrestrial world and its afflictions without any break of continuity between the macrocosm and the microcosm, the living and the non-living, the disease or affliction and its remedy. It is an example of the detailed working of the chemical philosophy as against the mechanical philosophy of the European modernity.

The principle of causality or effectiveness at work here is then not simply that of resemblances or similitude as established between the four or five separate items or distinctive features of anaemia, blood (which at the time was reckoned as an organ), the planet Mars, iron and chemical medicine put in a series by the principle of analogy, which will be reducible to a merely subjective association of ideas. It is rather that the *systems of differences* found among the various specific diseases, organs of the human body, the planets, the metals and among chemical compounds as medicines all formally resemble each other *and they also form a single unbroken fabric of existence* in a field of correlation, pairing and sharing, e.g. by the principle of homology or symmetry. If one were now to ask Paracelsus or a Paracelsus scholar to explain the *mechanism of interaction* or of continuity and discontinuity in this chain or fabric of being the answer would be probably in terms of the 'semina' which are supposed to be invisible and all-pervasive forces of the universe that act to produce form, govern function and give things their respective generic and individual properties and qualities rather like the phenomenon of electromagnetism as against gravitation. Paracelsus had derived this truth, method and praxis from the relation of the whole and the part (structure) or of self, the world and the other (discourse) as well as from the archetypal symmetry, equivalence, correspondence and complementarity of the macrocosm and the microcosm, thereby leading him to understand the sympathy as well as the antipathy of all parts of creation in mutual interaction, attraction and repulsion. The formula or the principle of efficacy *similia similibus curentur* (like cures like, now called homeopathy) was thus superadded by Paracelsian medicine to the *contraria contrariis curantur* (unlike cures unlike, curing by contraries, allopathy) of Galen and the Greek medicine. For this reason we have in history one and the same Paracelsus being claimed as the original founder of modern chemical medicine or iatrochemistry as well as of the science of homeopathy in Europe, although the two aspects have once again become wholly dualistically separated, divided and antagonistic since his day.

It is still to be determined how much of all this Hermetic corpus of medicine and natural science or of cosmology and anthropology is connected with Paracelsus' non-dualist eucharistology, a part of which exists still in rare manuscript texts only (e.g. the early seventeenth century text *De coena Domini*, MS Harley 508, British Library, London). It seems that the true object of the Eucharist—and of the other Christian sacraments—is *pharmakon athanasias*, to impart immortal heavenly medicine to the earthly and the mortal, the sacramental body to the elemental body, showing the enfleshed spirit in and through every human person. In his commentary on the gospel of Matthew written in the early 1530s, Paracelsus argues that Abraham was born of another order or limbus, separate from the blood of Adam and Eve, so that the virgin Mary and hence through her Jesus and all those who partook of his body and blood in the Eucharist would have a different nature, the special Christian power born of the new limbus in Christ. Thus participation in the rites produces the real creaturely transfer of the communicant from the 'cagastric' body (mortal) to the 'iliastric' body (transfigured), from the old life into the new birth, health, sanity, light and immortality (longevity).

The best single example of the above principles of reflection, symmetry, equivalence, correspondence and complementarity or of the mystery of correlation, pairing and sharing that I have been able to find in the Hermetic tradition is connected with this Paracelsian mode of thought, code of conduct and principle of organization. It illustrates the double figure of the macrocosm and the microcosm, the figure of the world (left) and the figure of man (right), as a representation of their inverse relation of complementarity, which is here implicitly defined as each of them being made up of the same or analogous elements but occurring in the reverse order of equivalence or correspondence so that the two together form a single symmetrical whole (fig. 1). This illustration is taken from the early seventeenth century text of *Introductio hominis*, a 'short guide to a godly Christian life', which itself forms a part of the *Philosophia mystica* (Neustadt, 1618), a compilation of eleven religious treatises in German by Paracelsus (1493–1541) and his follower in religion Valentin Weigel (1533–88). The twentieth century historian of science and medicine, Pagel, who has reproduced this unusual figure has already duly noted the 'inversion of order' between the pair of terms, but he shows little or no interest in pursuing its underlying principles of correspondence and complementarity.

> The divine Light and Spirit (*spiraculum vitae*) dwells *outside* and above the elemental world, but occupies the *innermost* centre of man. *Outside* the latter there is the darkness of the earth [the devil] and the abyss of hell—which in turn occupy the *centre* of the cosmos.
>
> The same inversion of order applies to the concentric layers of the [four] elements: [a] 'spiritual water' in the greater world corresponding to *fire* and mind in man; [b] 'fire' corresponding to *air* and soul, sidereal spirits—including the *evestrum* through which man receives messages from the outside world and can make forecasts concerning cosmic events—and reason; [c] 'air' corresponding to *water* and blood ['for no blood is without water']; [d] 'elemental water' corresponding to *earth* and flesh (Pagel 1958: 119).

The two corresponding and complementary figures, *figura mundi* (the cosmos) and *figura hominis* (man), are both conceived in the shape of two equal and equally incomplete sets of four concentric semicircles but centred respectively on the dark of the earth and second on divinity, the light or spirit. The

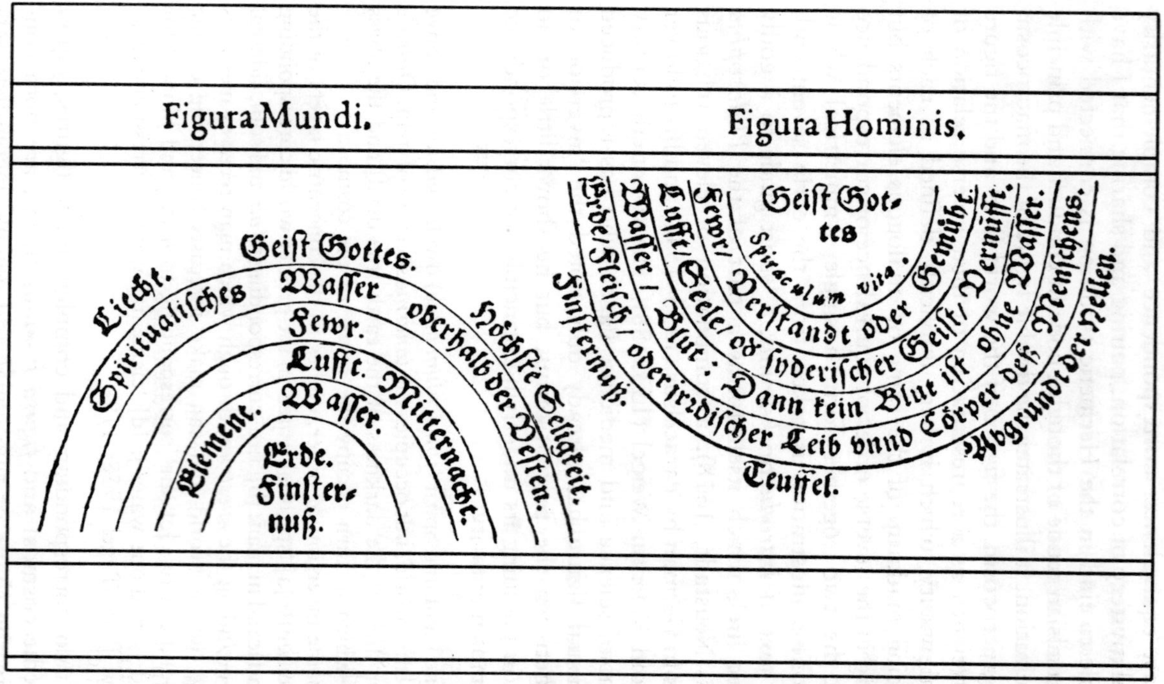

FIG. 1: Macrocosm and microcosm in the Hermetic tradition: figure of world (left) and figure of man (right), showing their relation of incompleteness, reflection and complementarity

Source: Philosophia mystica *in German by Paracelsus and Weigel. Neustadt: Lucas Jennis, 1618*

movement from its lower dark horizon to its upper light horizon in the case of the macrocosm means a movement of expansion or exhalation going outwards from its inside to its outside, whereas conversely or inversely the same journey in the case of the microcosm means a contraction or inhalation going inwards from the outside to the inside of man. The first journey is shown by the great world rising above the lower dark horizon from the centre of the earth through 'elemental water', air, fire and 'spiritual water' to the sphere of spirit and light, its upper horizon. The figure of man, on the other hand, is shown inverted below its upper light horizon proceeding downward/outward from the inner light of the spirit through its double corresponding order of elements, the first two being mental and the last two being bodily—i.e. fire (mind), air (soul, sidereal spirit and reason), water (blood) and earth (flesh)—to the outer darkness, the devil and the abyss of hell. This double or dual nature of man is, as we have already noted, a general characteristic of Hermetic science, religion and philosophy, his one side being more soul than body and the other being more body than soul, the unity in the duality.

Using a purely geometrical or topological language, the macrocosm and the microcosm are here represented as two equal half-figures but they are not shown as being situated exactly opposite to each other so that the two aspects are each of them separated and incomplete but together they constitute a single symmetrical whole, like any pair of terms of the unity in the duality. Thus the operations of transformation that are required to be *performed by the subject* in order to make (say) the first half-figure go into the second half-figure would be (a) to reflect the first half-figure of the macrocosm in its own dark horizon used as a mirror or alternatively to give it a self-rotation through a half-circle or 180° and then (b) to slide it along an upward diagonal glide axis or simple axis of translation as mathematicians call it, stretching from the lower earthly centre (left) to the higher spiritual and lighted centre of man (right). Alternatively, one has to imagine or perform the same two operations of reflection or self-rotation and then translation vice versa on the right hand half-figure of the microcosm so as to obtain the left hand half-figure of the macrocosm. It is only in this sense, I want to say, of the principles of homology or symmetry, polarity, equivalence or correspondence and complementarity *and* with a certain defined space or distance left between the pair of two terms that we should understand the often-quoted Hermetic formula that 'the world is a great man and man is a little world'. It is not a matter, therefore, simply and only of analogy or similitude plus a separateness of place and difference of size or scale, as usually implied by the many figurative representations of the period from Paracelsus to Fludd, where the one term of the macrocosm and the microcosm is shown as being just the same as the other only being writ larger or smaller, which would surely make one or the other of them redundant in the total cosmological economy!

The macrocosm and the microcosm are surely each of them incomplete without the other, and *therefore* the two of them together as a unified pair form a single unbroken chain of being whose links of interaction in space and time, succession or simultaneity, are various parts that are either (a) separate but similar (unity of origin) or else (b) different but together (unity of goal), the single interwoven fabric of the cosmos.

The critical defence of Paracelsianism proffered early by Daniel Sennert (Wittenberg, 1619) at the beginning of the official scientific revolution already alleges that Paracelsus attempted to replace human reason and experiment, the two time-honoured methods of science, by his vision of similitudes in nature, whereas what mattered was the search for causes (and mechanisms). It is evident that from the beginning of the early modern period (c. 1500) the cosmological theory of correspondence and complementarity is all too often misunderstood one-sidedly by its defenders as well as its critics to rest upon similitudes and the relation of analogy (parallelism) to the virtual exclusion of differences and the relation of homology (e.g. polarity). This widespread tendency badly needs to be corrected as I have argued by amplifying the principles of symmetry and reflection, correspondence and complementarity, which are freely available everywhere but little used at any rate from Dante onward except by followers of the Hermetic cosmology from the Paracelsian reform to Goethe as a scientist, representing the other self of the European modernity.

IV

CONCLUSION: MALE AND FEMALE

Auld Nature...
Her 'prentice han' she tried on man,
An' then she made the lasses, O.

Robert Burns, 1759–96

If and where it exists and lives today, the Hermetic tradition is somewhat deficient in its sense of history and of society, perhaps just because it had remained underground for so long. It appears to be lacking in the Judeo-Christian concept of the congregation or the community of the saved as much as in the Christian forms and concepts of the saviour, the scripture and the sacraments (rituals). Its non-dualistic attitude to the conflict of immanence versus transcendence perhaps blunts the absolute primacy of the principles of the fatherhood of God and the brotherhood of man, the community of humanity. It might be said also against Hermetism that its emphasis on the unity of God, man and nature is much too kind to non-human nature, forgetting the aspect of nature red in tooth and claw, and not mindful enough of fearing God as well as loving one's human neighbour. Its strongest point then would still be left, I suppose, as the celebration of the combination of equality with difference, e.g. in the human species' relation of male and female—if not as before in the relation of the macrocosm and the microcosm.

For my final example, therefore, I select a recent paper by Laqueur (1987) on the politics of reproductive biology to show the perennial underground life of the Hermetic tradition, which may surface in the oddest places. The paper is written to the effect that 'at one time male and female bodies were regarded as hierarchically, that is vertically, ordered and that at another time they came to be regarded as horizontally ordered, as opposites, as incommensurable.' In Europe the machinery of sex was an 'exercise in topology', Laqueur writes, at any rate in its Galenic structural explanation as against the tradition of Aristotle:

Turn outward the woman's, turn inward, so to speak, and fold double the man's, and you will find the same in both in every respect.

Think first please, of the man's [external genitalia] turned in and extending inward between the rectum and the bladder. If this should happen, the scrotum would necessarily take the place of the uterus with the testes lying outside, next to it on either side.

Think too, please, of the converse, the uterus turned outward and projecting. Would not the testes [the old Latin name for the woman's ovaries] then necessarily be inside it? Would it not contain them like a scrotum? Would not the neck [the cervix], hitherto concealed inside the perineum but now pendant, be made into the male member?

You could not find a single male part left over that had not simply changed its position.

It is to be remembered that this picture was consistent with the Galenic-Hippocratic two-seed model of conception in which a woman, contra Aristotle, actually 'seminated' at the peak of her sexual raptures, although the female penis (or woman's yard, Engl. vernacular) was not discovered by official medical Europe until the time of Renaldo Colombo, Vesalius' successor in the chair at Padua in the sixteenth century (*De re anatomica*, 1572).

Thus, until the very end of the seventeenth century there seemed no difficulty in holding [a] that women had an organ homologous, through topological inversion, to the penis inside their bodies, the vagina, and [b] another one morphologically homologous to the penis, outside [the body], the clitoris.

The making of the modern woman's body would thus appear to be a fairly recent phenomenon, and this will have to serve for our present conclusion. Thus the European modernity has not solved the mysteries of cosmological symmetry and homology or of (say) pairing and sharing, the warp and the weft, sympathy and antipathy, the origin and the goal; it has simply chosen not to pose them as unaskable questions.

With these interpretations [c. 1900] of spontaneous ovulation the old physiology of pleasure and the old anatomy of sexual homologies were definitively dead. The ovary, whose distinction from the male testes had only been recognized a century earlier, became the driving force of the whole female economy, with menstruation the outward sign of its awesome power (Laqueur 1987: 3ff, 7, 14, 28).

2

The Body of Christ and the Origin of Modernity

> More than ever before, Catholic theology at this time [AD 1190–1250] was a statement about light. In order to combat the temptations offered by the Albigensian heresy, the best of the theologians harked back to the system of hierarchies described by Dionysius the Areopagite, and sought to shore up that monument with sturdier reasoning and enrich it with the progress that had been made in physics. Robert Grosseteste, who founded the first schools in Oxford, the nucleus of the future university, read Greek fluently; he was familiar with Ptolemy and the new astronomy and with the Arabs' scientific commentaries on Aristotle's *Treatise on the Heavens*. In Grosseteste's philosophy, God was still light, and the universe was a luminous sphere that radiated outward from a central source into the three dimensions of space. All of human knowledge stemmed from a spiritual irradiance of uncreated light. Were it not that sin makes the body opaque, the soul would be able to perceive the blaze of divine love directly. It was in the body of Christ, both God and man, that the corporeal universe and the spiritual universe reverted to their original oneness. Hence Jesus was designated—and, along with him, each of the cathedrals, as so many symbols of Jesus—as the center from which all things proceeded and received their light: the Trinity, the Word incarnate, the Church, mankind in general, and all of creation.
>
> Duby 1981: 147

I

THE PROBLEM OF THE ORIGIN OF MODERNITY

As is well known, after its first thousand years Christendom came to accept the great schism between the western and the eastern, the Latin and the Greek, churches as permanent (AD 1054), and their mutual excommunications were not retracted until 1965. The chronology of the other great break in the unity of Christendom, i.e. the division marked by the Reformation and the counter-Reformation, is perhaps best encapsulated by the Council of Trent (1545–63). This was probably the longest council ever to be convened in Christianity and it produced the modern Roman Catholic church and its doctrine, on the one hand, at the very time when the Protestant churches were being founded by the reformers, Luther, Zwingli and Calvin, on the other hand. If the controversies of the first period were concerned with (a) the relations of Jesus Christ to God and to man, I hope to show that the controversies of the new modern schism were concerned with (b) the relations of Jesus Christ to man and to nature. So opened in effect in the mid-sixteenth century the modern chapter of man and nature as well as of natural science, owing as much (or more) to the Reformation as it did to the Renaissance.

The thesis is that the manifest focus of the Reformation and the counter-Reformation was upon the question of the liturgy, i.e. the mode of presence of divinity in Christian ritual, as much as upon questions of faith and the church hierarchy. The chief argument took place between the Pope and Luther and Zwingli as the three imagined parties. Their respective positions may be called transubstantiation, consubstantiation and non-substantiation, but they all implicitly made certain modern assumptions such as the mutual alienation of the truth (*veritas*) and the reality (*realitas*) or the spirit and the form. Altogether the debate marked the watershed between the medieval and the modern periods of Europe as representing two distinct world-views and life-worlds, whatever their respective varieties. It was given to Zwingli, if we single him out, to expressly formulate the new system of interrelations between God, man and nature, which in turn produced the new man and the new secular philosophy, the sciences of man and the sciences of nature.

In other words, what we call modernism or modernity as a faith, a regime and a system in the wide sense first arose in Europe, as it would arise anywhere, over the question of the place of God and the sacred axiology in the world-view and the life-world of man. Modern positivist dualism first attacked and successfully demolished the medieval Christian synthesis, transcendentalist and immanentist, in the sphere of religion, i.e. in relation to divinity or spirit and its symbols, and only then went on to unfold on its own the new relations, categories and attitudes of modern life and thought. The shortest way I know to define and explain the new regime or structure as a whole is to say that it rested on two fundamental presuppositions: (a) the dissociation and the autonomy of fact and value in the field of knowledge; and (b) the dissociation and the autonomy of lexical truth and applied praxis, or theory and technique or consciousness and conscience or belief and conduct, in the field of life. The interrelation of the fields of knowledge and human life, truth and method, within a still larger whole consisting of, say, God, man and nature, it left as an entirely open question on which its position was, as it were, perfectly agnostic. The implicit theory of modern practice and the practice of modern theory together built and unfolded a powerful new regime of science and culture, philosophy, politics, economics and ethics, which has lasted to this day.

Up to and partly including the period of the Renaissance, it seems that any European could pursue knowledge of the true, the good, the beautiful and the useful conjointly as *scientia*, as did Leonardo da Vinci (1452–1519) for instance, while relegating to the outer darkness, to witchcraft and the occult, knowledge of the false, the evil, the ugly and the useless. The first pursuit was holy and the second unholy, the first superordinate and the second subordinate, almost by definition. After the great schism of the Reformation and the counter-Reformation in the mid-sixteenth century, however, the order of priorities and distinctions between and within the two sets of terms was completely reversed. The student was now asked to determine first and foremost whether the object of knowledge fell within the competence of the mathematico-experimental sciences, the moral and political sciences or the arts, whether useful or ornamental. Since each one of the three great branches of knowledge and their corresponding practical arts purported to deal with separate and different questions autonomously, the determination of truth versus falsity or good versus bad or beautiful versus ugly or useful versus useless could take place only afterwards and secondarily. It followed logically that each one of these latter determinations in the arts and sciences, pure and applied, was to be done somehow independently of the others.

Now I think that the seemingly purely theological debate which took place at Marburg in AD 1529 over the nature of the Christian ritual sacraments between Luther and Zwingli, both of them apparently Protestant reformers, marked the final *inward* division between the old and the new forms in the history of European civilization. At the same time, the almost simultaneous *external* manifestation of modern western consciousness as well as conscience was marked by the so-called Age of Discovery when the Christian west cut itself off from the rest of the world in its own eyes and so was able to set out to discover, subordinate and conquer or to destroy the rest without fear of God. The relation between the outward manifestation and the inward manifestation of the new modern age was, as we shall see, precisely the kind of question that the modernist philosophy or regime was from the beginning determined not to face and solve.

It is well known how the term Protestant was first applied to those who favoured the cause of Luther and who protested against the intolerant majority decision at the second Diet of Speyer (1529) which revoked the earlier decision of the first Diet of Speyer allowing the church reformers a say in limited cases (1526). The Protestants then were all those who severed their connection with the Church of Rome at the time of the Reformation; and it is said that the essence of Protestantism was the acceptance by the individual Christian of his or her direct responsibility to God and to oneself rather than to the church, thus placing the local congregation of souls above the hierarchy. But this was not the whole story or its quintessence in my opinion. We should try and reformulate first the key question that drove the old Europe to the new schism, and the answers to which subsequently defined the Reformation and the counter-Reformation alike.

II

THE MARBURG COLLOQUY (1529)

In that historic and crucial year of 1529 the Lutheran prince, Philip of Hesse, had summoned both Luther and Zwingli to meet at his castle at Marburg in order to work out a Protestant consensus which he would back, but their public debate was wrecked by a fundamental disagreement over the truth and the reality of *the symbol* in relation to Christianity. The fourteen articles that were drawn up at Marburg recorded that the two parties had come to an agreement on every point except one, the mode of presence of Christ in their rituals. That is to say, the chief point of contention was: just how precisely had he lived and died and yet left for the faithful, through the ministry of his priests and in virtue of the true words said and repeated, his own real flesh and blood to be the meat and drink of Christians, to be fed upon under the appearance of a little bread and wine for the showing of the Lord's death until he come again? The Marburg colloquy foundered on this issue and it ended more or less with these words:

> We have not at this time come to an agreement as to whether the true body and blood of Christ are corporeally in the bread and wine.

I think that the key to the whole story of the new schism lies in discovering the original significance of these few words and specially the duality that they presumed (or newly established?) to exist between a spiritual or mystical presence and a corporeal or material presence. By stating this issue and forcing it into the terms of dualism, or perhaps more properly of double monism, Zwingli had discovered or invented the modern conception of time in which every event was either spiritual and mental or corporeal and material, but no event was or could be both at once, as in the Christian world-view and life-world of combined transcendentalism and immanentism. It was just as if Christianity, the religion of the saviour as the one and only mediator, the unity in the duality, had simply relapsed into a kind of Manichaeism, the religion of all duality without any mediation.

Firstly, I should say that Luther (1483–1546) was himself far from being the architect of modernity as some people would like to have him, but represented fairly closely the traditional interpretation of the Latin church. His famous earlier ninety-five theses already nailed to the door of the church at Wittenberg in 1517 were not articles of theological dissent but rather moral complaints against church abuses. Indeed, at Marburg Luther argued essentially that the sacred rituals of Christianity embodied the real and true point of intersection or 'consubstantiation' of divinity or spirit and the profane world and were effective in that capacity. Symbol, myth and ritual genuinely called forth or reconstituted *the real presence* of Jesus Christ's true body and blood in the Eucharist or the Lord's supper, the supreme Christian ritual. The means of communion with the risen Lord and the objective guarantee of the forgiveness of sins was the real presence of his body and blood, which were present not only to the eye of faith but truly 'in, with and under' the bread and the wine and really present even for the unbeliever.

Secondly, I would even go so far as to accept the modern view, although it seems to be a post-Reformation one and a reformulation, that Luther's doctrine of consubstantiation, i.e. the real substantial presence of the body and the blood together and along with the bread and the wine in the Eucharist, was halfway removed from the supposedly traditional doctrine of transubstantiation, i.e. the conversion of the whole substance of the bread and the wine into Christ's body and blood respectively, leaving only the appearance or 'accidents'. But I hold that the radical rupture in the terms of the debate was due chiefly to Zwingli, the so-called third man of the Reformation, rather than to Luther or Calvin, and that the Council of Trent in a way recognized the fact. Research should take interest, therefore, not so much in the question of the Roman Catholic versus the Protestant as in the difference among varieties of Catholicism and of Protestantism so as to explain the decisive break of Zwingli in 1529 with Luther *and* the Pope, who turned out to have so much in common from the new modern point of view.

Thirdly, Luther had consistently also held like the pre-Reformation churches that the sacraments were the individual's indispensable means of salvation along with his or her Christian faith and conscience. Faith, obtained through the grace of God, ritual and good works were together parts of a single religion and not three alternative possibilities of salvation as they were to become in Christianity later on.

From the point of view that was subsequently adopted by the Council of Trent (1545–63), which had apparently met to define and provide the Pope's answer to Luther, there were in retrospect two distinct rituals that were conjoined in the pre-Reformation churches. What was called the Eucharist had been in the medieval period at one and the same time both (a) the supreme sacrament of the Eucharist regarded as sharing in the Lord's last supper, and (b) the Mass regarded as being a true sacrifice participating in the original sacrifice of Christ on the cross. On the face of it, Luther himself had only wanted to restore 'the word' to its rightful place in the sacraments and so to remove the (priestly) 'superfluous miracles'; his avowed reform was aimed to preserve the Eucharist and its invisible saving grace as the supreme sacrament. Zwingli had, on the other hand, advisedly set out to attack the sacrificial ritual of the Mass. This difference of emphasis, the defence of (a) and the attack on (b), can be suggested to explain in some sense their difference of opinion, but I believe that the consequences of the debate itself were truly momentous for the world.

Zwingli (1484–1531), the reformer of Switzerland, had never enjoyed in his life the kind of secular support which lined up behind Luther in Germany, and Zwingli left behind no church in his name. His importance is dismissed very lightly by the sociologist Weber and even by the historian Troeltsch. Zwingli was to start with the enlightened child of the humanist Renaissance of Italy, and he was invited to the Swiss principality of Zurich in 1519. He there campaigned against the ritual of the Mass until he got the council of Zurich to abolish it in 1525. To properly assess this action alone it is necessary to recall that to medieval Europeans the Mass had been the chief mystery of the church, regarded as a holy drama both rehearsing and exhibiting again and again the story of the Lord's passion 'till he come'. Jesus Christ had offered himself on the historical cross and he again offered himself daily in the Mass before the people's very eyes and in their lives. Already in 1264 Aquinas had explained in the presence of the Pope that the sacrament of faith affected the body as well as the soul: it restored health and vigour to the sick and the weak as well as giving grace and virtue; it was the consolation of the dead as well as the completion of Christ's mystical body, the church. But Zwingli was out to overthrow all such 'priestly miracles' and would have nothing to do with any thought or practice of a repetition, renewal, participation in or continuation of the sacrifice on the cross.

Zwingli constantly stressed rather the social side of Christianity, i.e. the relation of the sacraments to the community at the present time. The Eucharist was for him only the outward symbol, so to say, of the inward union and communion of all believers in Christ, a kind of renewed sign and pledge of common allegiance to Christ and membership in him. As against the medieval ritual sacrifice of the Mass, Zwingli went ahead and set the Eucharist in the form of the Lord's supper regarded simply as a commemoration, a memorial or a remembrance of Christ's sacrifice offered once and for all. The primary reference of the reformed Eucharist was to be the physical and historical death of Christ (the crucifixion) rather than the metaphysical event of the risen Christ glorified (the resurrection), and certainly not both at once. The bread and the wine were merely symbols and figurative signs, certainly not symptoms, of the broken body and the shed blood of Christ, at most only the metaphorical seals of this purely spiritual union, the life of Christ in us and ours in him, which was renewed by God.

So in 1527 when Luther published his tract to the effect 'That the words of Christ, *This is my body*, still stand firm', Zwingli at once replied with his tract 'That the words of Christ, *This is my body*, would eternally keep their ancient and sole meaning; and Martin Luther with his latest book has not made good his own and the Pope's interpretation.' There were other similar exchanges between the two men which are on the record of history. From the beginning of his career right up to the end Zwingli insisted that in the utterance 'This is my body' (*Hoc est corpus meum*) the existential word 'is' (*est*) was to be understood, not in a real, literal and corporeal sense, but only in a symbolical, historical or social sense (*significat, symbolum est* or *figura est*). In effect he insisted that in this sinful and fallen world the two opposed senses were to be regarded as mutually exclusive, separate and different. The literal reality (fact) was to be regarded as being only of the body and this world, while the spiritual truth (value) was to be regarded as being only of the soul or the mind. Dualism or double monism was to be permanently fixed in the world-view and the life-world of the modern age, which was thereby ushered in.

It had been all in vain for the mystic Eckhart, who incidentally belonged to the same Dominican order as Aquinas, to have argued in his defence some two centuries earlier that it was evident to him that the holy scriptures had to be explained and understood 'parabolically', as he put it, but that it was not any the less true on that account that they must also be explained and understood by the Christian literally and historically.

To sum up the terms of the debate, we may say that Zwingli completely denied at Marburg against Luther and the Pope the efficacy of traditional Christian rituals. Before the schismatic revolution, the whole visible cosmos was the body of God, more or less, and it might be viewed as such by students of nature. By the new definition, inward spiritual faith alone was Christian and effective in any real and true sense, not the symbol, image, myth, drama and ritual. The spirit spoke only to the spirit, Zwingli said, and the sole aim of the divine service was the preaching of the word of God. Therefore, the ritual of the church should give way to the word in the chapel. The ultimate mystery of Christianity still remained somewhere perhaps between the ideal form of God and the 'priesthood of all believers', i.e. the local congregation of worshippers, but it was now wholly removed and excluded from the individual Christian's world of sensory perception, objective experience and outward existence. For the unregenerate man of modernity who lived in a thoroughly dualist world, there would be no manifestation of God in space and time: 'He is appropriated by faith alone.'

Of course, I do not wish to imply that Luther and Zwingli held nothing in common that went against the Pope and the orthodox Church of Rome. There were other well-known contributions of Zwingli which can be said to be parallel to those of Luther, namely, his view that only the word of the Bible, which he was the first to take as his text and preach directly from in his day under the guidance of the Holy Spirit, possessed spiritual authority over the faithful as against the claims of the

Pope and the medieval church hierarchy, and his active moral attitude to the secular world of work, etc.

III

THE NEW RELIGION AND DUALISM OF MIND AND THE WORLD

I hold that the sum total of Zwingli's revolution was finally (a) to divide the world of spiritual truth and the inner light from the world of apparent reality and outer forms; and (b) to insist that the two spheres of existence and experience, the inner and the outer, never met for man in this as against the next world. The two spheres were to be regarded as separate *and* different until the last judgment of God, since there was no medium left for their interaction, and this would lead to the infinite regression of dualism.

No non-dual Christian unity of the truth and the reality, the spirit and the forms, was ever presently attainable in a sinful world even for a blessed moment. The new Christian would no longer seek to transcend the duality of the spiritual and the temporal through ritual and the church as in medieval Europe, but rather, if at all, through the word in the chapel and work in the world. This work alone would show in this life and world whether a man was predestined to be saved for the other world or not (Calvin). For the Puritans and the dissenters, the Baptists, Independents, Presbyterians and Methodists, the mediation of the church would give way to the mission of the chapel.

For the purity of his modernism and radical anti-ritualism over the question of the Eucharist, Zwingli thus ruined the chances of his political alliance with the Lutherans which Philip of Hesse had wanted in 1529. Later, Zwingli himself ventured into the field against those cantons of Switzerland which had remained faithful to the Church of Rome, and he died a martyr to the cause in the battle of Kappel in 1531.

The sacramental controversy over the real presence had shaken and split into two or more parts what had been, in the words of Thomas Aquinas, the central pillar of the church. What had been, again according to Thomas Aquinas, the greatest miracle that he ever wrought, i.e. the supreme Christian ritual, which was altogether one and indivisible, subjectively and objectively, so that the Lord was present whole, entire and perfect in every fragment of the symbolic species, was henceforward to be regarded as visible only to the eye of faith and only on Sundays. Zwingli had said clearly, when disputing against Bucer as well as Luther, that what was present in the Eucharist is only by true 'synecdoche', a kind of metaphor (*mens*, in the mind of the believer); and that what he denied is any real 'metonymy': you cannot bite, Zwingli said, the body of Christ between the teeth (*dentibus ter[r]i*).

We may now put it that the reformulated Roman Catholic doctrine of 'transubstantiation' as well as the Lutheran doctrine of 'consubstantiation' both equally presupposed two separate and distinct spheres of spiritual and mystical versus material and corporeal presences. We have attempted to show that this was the new framework of thought presupposed or introduced by Zwingli's argument which might be called by analogy the argument of 'non-substantiation'. The debate at Marburg marked the frontier in Christendom between the medieval and the modern as much as between the Roman Catholic and the Protestant because in theory and practice whichever one of the modern set of three paths you adopted, following the Pope, Luther or Zwingli, the non-dualist modes of combined transcendence and immanence, if they ever existed, had gone for good and all.

The Pope's new doctrine now wholly removed, just like the Protestants to whom it was a response, the pantheistic element from the nature of the world. The Protestants had not only rejected the doctrine of transubstantiation but provoked the Council of Trent to redefine away out of its medieval past what they had rejected, the unity of substance and accident. Either way, whether it was affirmed or rejected, the new ritual of transubstantiation was believed to be a kind of miraculous transference by God's agency between two separate planes and two different times, i.e. from the material and the present to the spiritual and the eternal; it was no longer a normal transaction of God and man in this world.

Firstly, the trinity of spirit, word and sign had finally parted company for man at Marburg in 1529; and myth or ritual, say, was no longer literally *and* symbolically real and true, like a symptom or a signature. This modern dualist separation has been called the dissociation of faith and sensibility or of faith and knowledge. Several western writers have said that its wide and deep repercussions went far beyond the Protestant reform of the Christian church, and that it helped to produce modern man and the modern world in relation to the new image of God. I think that it is worthwhile to add that Zwingli was the chief architect of the new schism and that Europe and the world followed Zwingli in the event. Zwingli, the reformer of Zurich, was in his system of thought the first philosopher as he was in his life and death the first martyr of the modern world.

Secondly, I do not agree with those who thought—and many still do—that, on the other hand, the real issue of Protestant modernism was the question of authority, and that it lay between Luther and his traditional church superiors in the Roman hierarchy, which is the issue that also came to a head at the second Diet of Speyer in the same year of 1529. It seems that in modern dualist thought one must make the choice and attribute priority either to the idealist struggle for truth (Marburg) or to the material struggle for power (Speyer). But I argue that this is precisely the mode of thought and experience newly introduced by Zwingli, and that the whole issue was and is better interpreted as being one about the power of truth (the ritual) or about the truth of power (the church) or about both at once. I think that the battle of the big battalions, the Pope's and Luther's, decided little in the end, and that the modern European revelation, which irradiated the Roman Catholic and the Protestant alike after the Council of Trent had implicitly adopted the new rules of debate, was vouchsafed whole to Zwingli, the apparently lone figure.

Thirdly, we must also observe here that the crucial thirty-nine articles of the Church of England which defined the Elizabethan church settlement (1562) were themselves Zwinglian or very nearly so on the nature of the supreme sacrament. Although some high-church men may still speak figuratively of the Eucharist as a 'sacrifice', the duly established twenty-eighth article of the Church of England lays down that

> The Body of Christ is given, taken and eaten in the Supper *only* after an heavenly and spiritual manner. And the mean(s) whereby the Body of Christ is received and eaten in the Supper is Faith.

Finally, the sovereign as defender of the faith and supreme head of the church also made a declaration to the effect

> That if any publick reader in either of Our universities or any head or master of a college or any other person respectively in either of them shall affix any new sense to any article... other than is already established in convocation with Our royal assent; he or they the offenders shall be liable to Our displeasure... and We will see there shall be due execution upon them.

This was said, demanded and commanded of the universities scarcely thirty years after the dissolution of the monasteries (1534–39), which had itself left the world of learning newly exposed to the power of the state in England. All members of the universities of Oxford and Cambridge, which were the only two in England for the three centuries that followed, were thus required for reasons of the state as well as the church to adhere to the thirty-nine articles, among other things. The university test act, which eventually abolished subscription to the articles, etc., as prerequisite to taking a degree, was not passed by Parliament in London until 1871.

The university man (and later woman) of England was in the modern period generally familiar with, even a participant in, the world of Zwingli and the Reformation as much as of Erasmus, say, and the Renaissance. I would say this on the basis of many years of what is called participant-observation of the British university system, even if I were not able to produce all the evidence or go into all the historical details. The sceptic is asked, not to accept this as his or her conclusion, but to consider that, if my general hypothesis of the gift of Zwingli and the new religion seems too wide to be tested, as it well might to some people, then the limited and concrete empirical object of investigation to be borne in mind should be *the modern university* and its new regime. In the nineteenth century Helmholtz, leader of the sciences of nature in Germany, had proudly boasted in his address on academic freedom that the same spirit which overthrew the yoke of the Church of Rome had also reorganized the modern German university. The first modern Protestant university of Europe was established at Marburg itself in 1527, soon to be followed by Koenigsberg (1544) and then Jena (1558). The reopened discussion at Marburg on the mode of presence of divinity in Christian ritual had reciprocally also implied a new dualist mode of presence for man in the world. This new dualist mode in its turn led to the new conception and organization of human knowledge in the modern European university, e.g. dividing the sciences of nature and the humanities of spirit.

(It is difficult for the outsider without a knowledge of Latin and German to follow the controversies of the Reformation and the counter-Reformation, but I have utilized the introductory articles in English on different aspects of the new schism available in the *Encyclopaedia of religion and ethics.* See specially under the heading, Eucharist, in Hastings 1912: V, 540–70.

I have also consulted the standard English editions of Thomas Aquinas and of Eckhart's articles of defence as selected representatives of the official and the unofficial medieval Christianity respectively. The last two quotations are taken from the current edition of the *Book of common prayer and liturgy*, which was first compiled in 1549 and started to take its present form in 1559.

According to the articles of Eckhart's defence [4: 14], it seems that the matrix of the Eucharist in his interpretation was made up by the elements and the equation, bread : wine :: flesh : blood :: body : soul. Each pair of these oppositions was 'united in being', and all three of them were unified by the authority of Christ [John 6: 56]. This is an example of the non-dualist unity of the spiritual and the corporeal elements in medievalism.

The sociology of Zwingli in English in a strictly professional sense seems to comprise only Birnbaum [1959], who knows his German but unhappily inclines towards the materialist interpretation of history in its most undialectical form.)

IV

SCIENCE UNDER THE MODERN REGIME

The system of the categories, relations and attitudes of life and thought that we call modern western civilization was fully manifested first, not so much in early capitalism and the Renaissance in Italy, but in the sphere of religion during the Reformation at Marburg and the counter-Reformation at Trent,

1529 to 1545. Its dualist logic as a whole then quickly invaded the other fields of European life and thought and restructured the relations between them, e.g. economics and politics, the sciences and the arts. In the beginning, as Durkheim the French sociologist wrote, everything is religious, and I only want to say the same of the European modernity, including its science. According to the method followed here, this is not a matter primarily of assessing the relative weights of causes, factors or features of the European modernity, but of suggesting an hypothesis on the basis of a discourse analysis of civilization as to exactly how the exercise of human free will, if there is any, had its widest, deepest and strongest repercussion or significance and effect in history.

I may somewhat simplify the structure of the modernist regime by saying that its two fundamental axes were the division between fact and value, on the one hand, whether or not value was further bifurcated into the instrumental versus the symbolic; and the division between lexis and praxis, theory and practice, on the other hand. These dividing lines were drawn hard and fast and they marked out the four quarters of modern western civilization to be occupied by the new science, technology, philosophy and politics, economics or ethics. My thesis is that this is the elementary structure of the modern *positivist* regime as a system, which produced the inner organization of modern western science, on the one hand, and its relations with the whole of the European modernity, on the other hand (fig. 2).

The old and/or perennial non-dualist Godly emphasis of truth and method was on the unity of conception that inwardly linked the truth of every part of man's estate with the reality of the whole externally as well as for the subject. The new dualist regime's emphasis was on the autonomy and independence of the separate and different sections, strata or segments of the cosmos, inner and outer, but chiefly outer. Each varied aspect and diverse part was to be comprehended and methodically brought under control separately and independently of the others, piece by piece. God might still exist and be active in the sub-field of the philosophy of religion, of course, but he had died out in three quarters of the European modernity, including its science of nature.

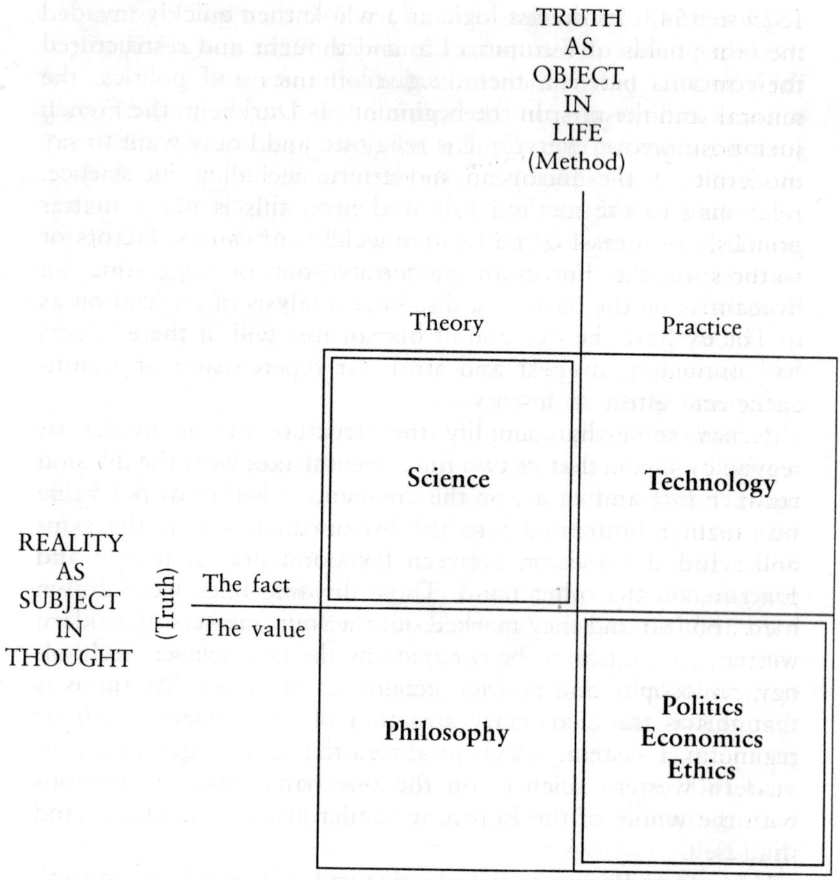

FIG. 2: Structure of the modernist regime

The dissolution of the unity of man's estate as a vision and a perspective to be realized in thought (truth) and somehow made concrete in life (method) led to the dissociation as well as the independence of its parts. Gloriously released from irksome mutual discipline under the ordered whole, the truth and the reality, thought and life, the inner and the outer, the truth of mind and the reality of the senses, simply flew apart, like God and the world, and renounced all necessary connection with each other (the horizontal and vertical axes on fig. 2). Their intrinsic dissociation left the new truths of modern science totally powerless to influence the ethical conduct of modern politics or economics and vice versa, just as surely as their mutual independence divided off the mindless praxis of amoral technicism from the spirit of philosophy (the diagonal juxtapositions on fig. 2). The inherent logic of the European modernity had subconsciously determined that the mutual relations and attitudes of its science, philosophy, technology and politics, economics or ethics should be arbitrary and extrinsic to the special merits and respective categories of each separate domain.

The external uses or abuses of modern scientific knowledge, its contingent applications, were supposed to constitute its sufficient justification for others, though not for the positivist elite, who inwardly knew *the truth* of what they were about as a whole but habitually disowned the misuse of any particular truth. For example, *a truth* discovered by the scientist was to be taught to others and supported by them, the non-elite, if and when it was applicable to the good, let us suppose, of Enlightenment democracy or socialism and economic development, but not necessarily because it was the truth. By the same token, the manifest *untruths* taught by religion were to be happily tolerated and even quietly encouraged by the new regime for their extrinsic value in politics and ethics, e.g. in promoting social stability and cohesion. I should say that by the year 1800 or so this was already the avowed opinion of Saint-Simon, the messiah in France who first put forward positivism as a self-conscious philosophy and a system—later wrongly attributed by his pupil Comte to himself.

Be that as it may, Zwingli's unregenerate man of modernity in a sinful world somehow came true and gave birth to a science or a series of sciences which took as its first premise and chief task (a) the systematic unfolding of the presumed *diversity, separateness* and *difference* of things in the world (heterogeneity). Positivist man and the new scientist could avowedly only understand the truth in its several parts or individual fragments. God alone knew the unity of all things and he was by now thoroughly inscrutable. The present policy of man the scientist was to divide and conquer the external world now and save the internal world, hopefully still intact, for the kingdom of God to come. It was as if a man were to insist in a fit of insanity or out of perversity that the outer circle of his horizon was visible and objective and therefore a fact, but that he himself, the knower, was merely a subjective value unfit to be included in his own scientific discourse. The only sight that modern western man ever had after that of the unity in variety that is truth would be through (b) finding or forcing merely *uniformity, similarity* and *resemblances* in nature or between nature and himself (homogeneity). We may call this the second premise and task of the new age.

In this positivist logic, either the first set of truths applied to the case or else the second set, but never both sets simultaneously to any event, system or context of situation. The positivist would not see the world according to the principle of unity in variety, but only in terms of the principle of homogeneity versus heterogeneity. It necessarily followed that the human mind should be applied to study all different things separately and only similar things together. Positivist dualism had wholly changed the relation of the whole and the part (structure) as well as the relation of the subject and the object (discourse), and non-dualism was ruled out of court in both cases, structure and discourse, truth and method.

There was almost no space left in the sciences of nature for the higher unity of mind and the world and of man's faculties of perception, cognition and volition, or for the dialectics of the subject and the object, or for the kind of unity in variety that transcends separateness and difference without neglecting them. All such modes of thought and experience were finally expelled or obliged to escape either into the thin air of religious idealism or into the marshlands of romantic art and irresponsibility, but in any case leaving behind the scientific firm ground to positivism. In the nineteenth century even the great labour of Marx would not restore the higher unity to the study of history, politics and economics as the sciences of man. In the twentieth century even the oft-cited great labour of Einstein would not restore the higher unity to the study of physics and mathematics among the sciences of nature, the macrocosm and the microcosm, in spite of his life-long search for a unified field theory of gravitation (Newton) and electromagnetism (Maxwell), not to mention also thermodynamics. Indeed the physicist Pauli once ironically said of Einstein's search that

'What God hath put asunder no man shall ever join.' I am unable to give any opinion about similar developments in chemistry or the biological sciences, ever since the cosmology of Paracelsus was taken to pieces, but one could name no less a person than Goethe as our witness in botany as well as optics.

Russell, who was the most able advocate in Britain of the use of the scientific method in philosophy as against what he called mystical, religious and ethical preoccupations, declared that in his opinion 'divide and conquer' was the maxim of success here as elsewhere (analysis). This led him to formulate the philosophy of logical atomism or absolute pluralism, as he called it, variety at the cost of unity. His senior and friend Whitehead, on the other hand, apparently took the other route and arrived at the organic conception of the universe as a whole, but this too was a vain effort at glueing together merely through links of simultaneity and succession the debris of the world left *after* the physicists had finished with it. Both men had neglected to start and to end with an adequate theory of the self and the other as well as of the world. In the twentieth century, therefore, neither philosophical atomism (analysis) nor philosophical organicism (synthesis), coexisting disjointed and separately, have dared to give any lessons to the special sciences which together represent aspects of the relation or its absence between the subject of study and the object of study, and this dualist regime of the European modernity is our theme.

Now let us return to consider the science of nature as a social institution of early modern Europe and ask the question, what was the west's real and true *social contract* of modernity? Altogether, I should say that science, religion and politics or the corporeal, the spiritual and the temporal had seemingly agreed under the new regime to go their separate and different ways. In England, it was Charles II who institutionalized the new arrangements in the middle of the seventeenth century and who set an example to other nations. In 1662, immediately upon the restoration of the monarchy, the private group of scientists already meeting in London, which had become known as the Invisible College for the cultivation of the new philosophy, was incorporated by royal charter to become the Royal Society of

London for 'the improvement of natural knowledge'. This autonomous establishment of the corporeal realm was just as important as, perhaps even more important than, the earlier medieval separation of the spiritual realm and the temporal realm, the church and the state. Anyhow, the latter separation was also renewed in form and effect in the same year by the Act of Uniformity, 1662. This marked the national separation and the persecution of the Nonconformists, the new Christians, who had severed connection with the established Church of England when it had accepted royal supremacy at the time of the Reformation.

On the one hand, King Charles II effectively bestowed the gift of autonomy on the cultivators of the new mechanical philosophy of nature; on the other hand, he implicitly confirmed the independence of the Puritans and dissenters of religion, the new Christians. In this way, the year 1662 set both the pattern for the internal organization and functional autonomy of modern western science and the pattern for its relations, or the apparent lack thereof, with politics as well as religion, discussion of either or both of which topics the Royal Society was enjoined to eschew.

Another century later and one could show that the intellect of the Industrial Revolution, 1760 to 1840, was produced by the mechanical institutes and the so-called dissenting academies of the Nonconformists, outside the two universities of England. The new inventors and captains of industry, from Watt to the dynasty of the Darbys, were not only all of them Protestants in relation to the Church of Rome but also dissenters in relation to the Church of England. In the structure of religion, society and culture, the Nonconformists stood precisely where Zwingli had stood in relation to Luther and the Pope. That is to say they were the men of work rather than ritual and of the chapel rather than the church or the state; and they were very often men who were dissenters by trade or occupation as well as by tradition. New developments of man in relation to his work were thus only to be expected of them; and such social origins were all of a piece with the whole logic of the European modernity in relation to science, religion and politics or the cycle of faith, knowledge and action.

V

POSTSCRIPT

A few years after my thesis was first published (1978), a new historical document came to light in Italy (1982) which has led Pietro Redondi to try and rewrite the history of the case of Galileo (1564–1642), both Christian heretic and leader of the modern scientific revolution of the mid-seventeenth century. It was generally supposed until now that Galileo was condemned by the holy office at his trial in Rome (1633) for upholding Copernicanism, i.e. the astronomical theory that puts the sun rather than the earth at the centre of the solar system. Now Redondi argues that Galileo had made in effect a secret deal with the authorities to escape the worst, so that he should be held guilty of a minor heresy (heliocentrism) and thus avoid the most dreadful charge and scandal of a major heresy, i.e. his challenge to the physics of the Eucharist.

Thus the trouble lay with Galileo's physics in the first instance and not with his astronomy; and mid-seventeenth century Roman Catholic theology could live with heliocentrism and the Copernican system rather more readily than with a physics whose atomism appeared to threaten the chief dogma of the Eucharist (transubstantiation). The newly discovered manuscript document is only about three pages long, a logically skilled but anonymous denunciation secretly lodged with the holy office in 1624 arguing that the Democritus-like atomism of Galileo's *Il Saggiatore* (1623) in effect brings into question 'the existence of the accidents of the bread and wine which in the Most Holy Sacrament are separated from their substance'. We are fully justified, therefore, in our experiment to examine the new science of nature as well as the new religion of modernity through the schismatic Eucharist controversy over the real and the true presence(s).

> Mock on, mock on, Voltaire, Rousseau,...
> The atoms of Democritus
> And Newton's particles of light
> Are sands upon the Red Sea shore,
> Where Israel's tents do shine so bright.
>
> William Blake, 1757–1827

3

The Other Science of Nature in Europe

It is true [a] that the kingdom of nature must be helpful to the kingdom of grace; but as everything is connected in God's great design, we must believe [b] that the kingdom of grace is also in some way fitted to the kingdom of nature, in such a manner that this keeps the greatest order [unity] and beauty, so as to render the whole composed of both [grace and nature] the most perfect possible.

Leibniz 1900: 298

The part played by measurement and quantity in science is very great, but is, I think, sometimes overestimated. Mathematical technique is powerful, and men of science are naturally anxious to be able to apply it whenever possible; but a law may be quite scientific without being quantitative.... We must, therefore, in dealing with such a matter as animal behaviour, be content in the meantime with qualitative laws [e.g. Pavlov's laws of the conditioned reflex] which are none the less scientific for not being quantitative.

Russell 1949: 67f

I

THE PROBLEM AND THE FRAMEWORK

The discourse of nature uttered through man or what we call science is not spontaneous and self-explanatory, in my view, whatever the ruling positivist assumption might be, but requires the discourse of culture and politics, i.e. knowledge of the culture/politics of the discourse spoken by man to man or to himself, in order to complete it. Therefore, we are proceeding here on the assumption that, apart from or along with its manifest content, the manner of formulating a question and correspondingly the *form* of the true answer expected from nature and the world, really help to define and indeed constitute the scientific and cultural/political universe in which a particular age, nation and class or community thinks and lives. To make such a large question somehow tractable for investigation is therefore also to define a particular viewpoint towards the history of the world. If we can possibly come to grips with these questions at all in India today, we shall have to make up our own minds independently, among other things, as to what had happened and how in the European modernity, which constitutes the chief single source of all our imports in the world of knowledge, power and culture. One can then begin to decide whether there exist perhaps other non-standard or alternative traditions and underground tendencies which could redefine the official European science or history, politics and economics or ethics handed down to us.

One possible hypothetical framework for such independent research would be to subdivide the history of the European modernity conveniently into three periods, stages or phases. Firstly its 'period of creation' or the Copernican/Galilean Age (1500–1650) to a consideration of which I shall return in a moment. Secondly its 'period of institutionalization' (1650–1800) when, for example, scientific laboratories and journals and new syllabi and textbooks for pedagogy really got firmly established in Britain, France and Germany. Thirdly the 'period of diffusion' (1800–1950) when this pattern of culture, institutional social organization and mode of thought were developed and spread throughout the world under European rule, supervision and example. We need not here at this moment go into the question of what further dimensions were added to the world system of the European modernity by the accession of the new territories of America, Russia and thirdly Japan.

One may look then for certain neglected symbols and events, texts, movements and persons from each period through which or whom one could perhaps discover and demonstrate the alternative tendencies that have been hitherto ignored by the established official histories of universal knowledge. For instance, instead of only re-studying Copernicus, I had looked for someone at the moment of change around 1500 who would be equally modern and anti-traditional or reformist but perhaps in some alternative sense, i.e. closely linked not only to the Renaissance in Italy but also to what has been called the movement of the radical Reformation, the Nonconformists or the left wing of the Reformation as well as to the great peasants' revolt in German-speaking Europe north of the Alps. It turned out that, as was said in chapter 1, the perfect candidate was Paracelsus (1493–1541), who is already known to be important in the history of chemistry and of medicine and also heavily involved in the history of magic, in which interest has latterly revived, but whose importance in the histories of religion and politics is yet to be fully discovered and assessed.

I now believe that it would be indeed possible to show that there was a Paracelsian revolution (the chemical philosophy) just as much as there was a Copernican revolution (the mechanical philosophy), but that the first was stalled and soon went underground probably because it was connected at birth with the politics of the (unsuccessful) peasant wars in Germany, and with a firm ideology and practice of non-violence right in the middle of the Wars of Religion of the early modern period. To put the question another way, the radical European underground was not just a political movement of the first period; it had equally a scientific aspect and a religious aspect, which have been neglected. In the new Christianity, people with a mode of thought and experience like Paracelsus were represented by the Anabaptists, for example, and they were soon officially and forcibly suppressed, often by horrible means even beyond the fire and the sword, although they behaved completely non-violently and believed in non-violence as a creed. Incidentally, Paracelsus was well regarded, translated and published on medicine, science and alchemy in England under the Puritan revolution and the Commonwealth (1649–60), although where exactly his religious influence and political example stood during this 'world turned upside down' period is not yet known.

Similarly, in the second period that we have distinguished around 1800, the Paracelsus method and praxis were revived and updated from the radical German underground through the philosophy of nature and the scientific writings of Goethe, as we shall try to show in the present chapter. Some of his themes were subsequently developed, so to say, from the right by Hegel's *Philosophy of nature* (c. 1830) as well as from the left by Engels' *Dialectics of nature* (c. 1880). Both of them have been of course systematically ignored and/or denigrated as merely contributions of metaphysics (the right) or of politics (the left), and therefore held to constitute non-science as judged by apologists of the official positivist-dualist methodology of Copernicus, Galileo and Newton. I will attempt to answer the question posed at the beginning of this investigation regarding the possibility of a different non-dualist modernity or perhaps post-modernity specially on the grounds of science as well as of philosophy, holding religion and politics in reserve for later. It is to be urged in defence of Goethe that, in his semiological theory of nature or the science of the symbol, he was being in effect neither a materialist nor a simple idealist, the modern poet who had strayed into the wrong field by mistake, but pursuing in his secretly anti-modern way something like the unity in variety of God, man and the world. We will try and illustrate this hypothesis with some technical experimental detail rather than only in general philosophical terms.

About the fundamental phenomenon of colour as seen respectively by Newton and Goethe, even at the latest international conference held by the Goethe institute on the theory of colours at Weimar, 1998, the participants appeared to be divided on the question of representation. The technical aspect of the question is as to whether (a) the familiar Newton representation of the phenomenon (only light rays) is objective and so to be projected on the screen, putting the emphasis on the refrangibility rather than on the hues of colour, the violet being refracted the most and the red the least, while (b) the corresponding picture of Goethe centres on the eye of the subject and therefore like the image of the rainbow cannot itself be simply projected on to a screen, and whether (c) there can be some non-dualist, objective plus subjective, reconciliation yielding the co-production of the subject and the object, involving the sun and the rain cloud as much as the human eye. The resultant unity in the duality might then indeed emerge as the common semiological language for science, art and religion, if not also politics and economics, making of them not a unilinear spectrum but a circle of cosmological as well as anthropological light and life, *zoe kai phos*, as a follower of the Hermetic tradition would say from the other self of the European modernity.

Newton's first paper of 1672 reduced the *difference of hue* between diverse sorts of light to the 'inequality of refractions' of various corpuscular rays. The best that you might get today is the admission of one set of hues for light (blue, green, red: physics) and another for pigments (blue, red, yellow: chemistry), etc., without any explanation of the trans-systems unity as well as the variety of the science of colour, chromatology.

II

AT THE FRONTIER: THE HUMAN BODY

> Nature, gutted of gods,
> Bent, like a thrall, to gravity's ponderous law,
> The dead beat of the pendulum clock.
>
> Friedrich Schiller, 1759–1805

As is well known, J. W. von Goethe (1749–1832) was a versatile universal genius, and apart from being a great German and European poet spent a large amount of his time and effort on methodical scientific work in optics and botany, geology, etc. Now the official science of the European modernity has responded to this by classifying and dividing Goethe's science of nature into two parts. (a) Natural history, the contribution that is classified as being merely empirical and 'descriptive', chiefly his botany and the morphology of plants, and that has been eventually absorbed into the official scientific 'mainstream' of course minus his so-called organic or vitalist philosophy. (b) Natural philosophy, the contribution that Goethe would himself classify as specifically physics, chiefly his studies on physical optics and the theory of colours, which are nowadays considered mostly wrong in fact and wrong-headed in theory and method. This latter work is supposed by its kindest critics of the modernist persuasion to be of interest solely to people who are really only concerned with his literary work and imagination or intuition. It was all only a mistaken excursion of his artistic genius into the wrong field of knowledge: a man of the arts, which are products of the right hemisphere of the human brain, had somehow wandered into the sciences, which are the work of the left hemisphere of the brain.

But let us, at any rate for the sake of the present experiment, try and take Goethe's work on optics seriously as physics, part of a theory of nature, including human nature, and not only as the product of a misapplied psychology of literature, metaphor or poetry. Goethe himself estimated that there had been greater poets than himself in the glorious past, there were equally great poets living in his day, like his friend Schiller, and no doubt there would be still greater poets to come in the future; but that he was the only man of his entire century who knew the truth about colour and yet no one would recognize his superiority to Newton, who was wrongly holding the field. I propose to take Goethe's self-assessment seriously and invite you to see where it might possibly lead us—back to the occult philosophy as the modernists say they fear or forward to an alternative non-dual science?

Those of us who are of the British empiricist, inductive and experimental habit of mind might like to start by setting up the two experiments and verify the two results, the Newton spectrum and the Goethe spectrum, with single fine light and dark vertical bands on inverse contrasted backgrounds viewed through a prism, both spectra reduced with the fewest possible colour bands obtainable, as shown in fig. 3. The Newtonian view is of course already familiar to everyone from our official school physics. If there were to be a simple polemical exchange between the two different views represented, perhaps it would be with the Newtonian claiming that the Goethe spectrum is really only two elementary Newton spectra muddled up by the eye in perception, and so not really a genuine physical phenomenon by itself which would require it to be necessarily projected on a screen independently of the observer. The Goethean would possibly return the compliment and make precisely the same charge in reverse, namely that the Newton spectrum is really

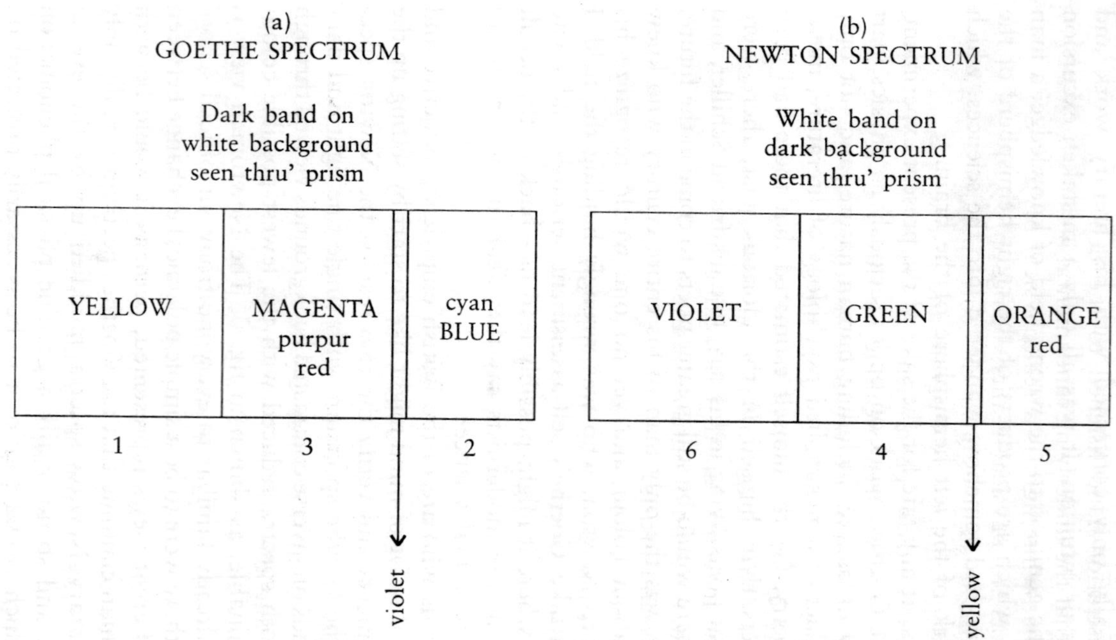

FIG. 3: Colours of the light and the dark

a complex mixture of two elementary Goethe spectra produced by their overlapping, addition or subtraction. If one watched the demonstrations and listened to the debate carefully, one would see that the underlying fundamental belief of the Newton school is that white light is a real physical phenomenon while darkness has no physical reality, being only a psychological phenomenon; it assumes also the corresponding dualism of physics and psychology as separate and distinct studies of the two different kinds of phenomena. A single fine vertical light or white band when seen breaking up through the prism into a spectrum of its monochromatic constituent rays is a natural physical phenomenon falling within the study of physics, while the Goethe view of its apparent inverse, the dark band similarly viewed on a white background through a prism and incidentally showing the magenta hue that is lacking in both Newton and the rainbow, is a kind of optical illusion useful only for artisans, biochemists and artists.

So those people who are by temperament or training of the opposite French, deductive or philosophical habit of mind would be better advised to start with the rationalist's question as to whether the two different spectra of Goethe and Newton are both only 'models of reality', right or wrong, or are perhaps also to be classed as models of our knowledge of reality, 'models of truth'? Or can there be still a third possible solution of mediation positing a non-dualist relation between the truth and the reality as the object of science?

The respective experimental and philosophical discourses will help us to explain that what Goethe truly wanted to demonstrate was that the light and the dark are not two dualistically separate and different realities (chalk and cheese) or even unilinear opposites (head and tail or north and south) in the way that they had been presumed to be hitherto. The phenomenon of the light and the phenomenon of the dark are both simultaneously physical and psychological, like the phenomenon of the rainbow, when seen from the single 'symbolical' view of (say) physiology and the human body or of the human senses and the imagination as well as reason. To make this representation at all intelligible, if not to prove it experimentally today, one has to assume I think that the underlying structure of Goethe's universe of discourse or cosmology is as follows. (a) The whole sensible world, *without the observer*, is divided into three parts: the medium, which is any kind of matter, solid, liquid or gas, that fills space, and another part that itself consists of two segments, the light and the dark, each of which is the condition of existence for the other, like the crest and the trough of a single wave or like motion in the middle and its apparent absence at the two ends of a swinging pendulum. (b) The whole intelligible world, *including the observer*, is perceived or assumed to be interrelated I think in this sort of way, i.e. understanding the unity and the struggle of opposites as somehow mediated and non-linear, occurring like the warp and the weft of a single woven fabric, as shown in fig. 4.

Thus we are asked to imagine the eye of the participant-observer as situated at two different positions successively—the first where it looks at the light through a darkened medium (yellow, developing to yellow-red and ruby), and the other where it looks conversely at the dark through a lightened

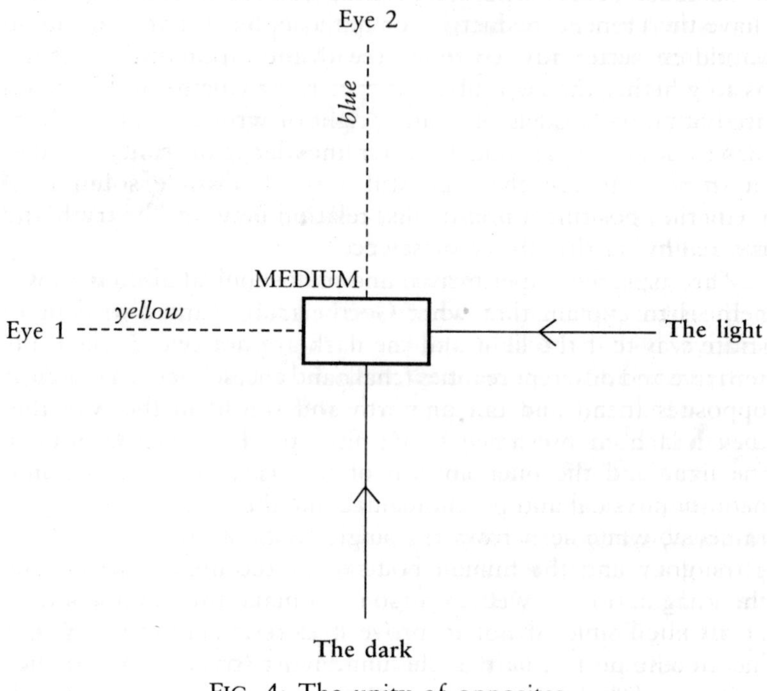

FIG. 4: The unity of opposites

medium (blue, developing to blue-red or violet). There is no experiment or argument in Goethe's *Theory of colours* (English trans. 1840) which would assume or imply that the light and the dark affect each other directly, physically or causally: this inference would be a simple misreading of the text. Goethe's experiments and argument are rather to the effect (a) that the light and the dark are truly separated and exist in different dimensions, if only as presence and absence, but that they conjointly and simultaneously affect the material medium defined as that third thing which functions in its turn effectively both to hold them apart and to bring them together according to the context of situation; and (b) that it is their respective threefold existence and mutual interaction in the roles of correlation, opposition and mediation that produces the effects on the eye which we finally call the phenomena of colour. This explanation and interpretation of mine are also borne out by some epigrammatic poems, c. 1810, that Goethe wrote at about the same time that his book on colour was published in German. I have therefore represented the phenomenon of colour in fig. 4 diagrammatically by showing the 'light' and the 'dark' as existing simultaneously along two perpendicular axes, and so intersecting or interacting rather like the foreground and the background only in relation to the medium and through it the eye of the participant-observer.

Goethe says quite plainly, and I follow him, that the two basic physical hues or colours are the yellow and the blue, howsoever they are produced. Eye 1 will thus see yellow because it looks at the light through the darkened medium, and Eye 2 will perceive blue because it looks in the opposite situation or configuration, i.e. at the dark through a lightened medium. I suppose that this also explains why we commonly speak in the English language of the yellow press, which is a medium that obscures the light, and also of blue movies, which is a medium that illumines the dark. The meaning and effect of all this is that the sensible/intelligible world opens and discloses the structure or the grammar of its whole system of discourse to the human eye or the sense of vision through the language of colour, which is not to be dismissed merely as a secondary quality of the world. The Goethe project here is the non-dualist Neoplatonist one of combining the human science of nature (the external world) with the natural science of man (the self) from a single 'symbolical' view with the human subject at its centre in relation to the reality of the senses, imagination and reason, which I shall try and define further.

It will be quite apparent by now that in this exposition Goethe had in fact taken the experimental results of Newton's prismatic analysis, where the colours (rays) are laid out from the violet to the red ends along a unilinear spectrum and quantitatively according to their respective wavelengths and/or frequency, and then transformed the series to curve back upon itself so that we can notice that its two ends, violet and red, have some quality like redness in common, i.e. maintaining the purely quantitative sequence but trying to add to it the human perception of qualities. Goethe in fact rejoined these two ends of the Newton series by introducing the new colour or hue that

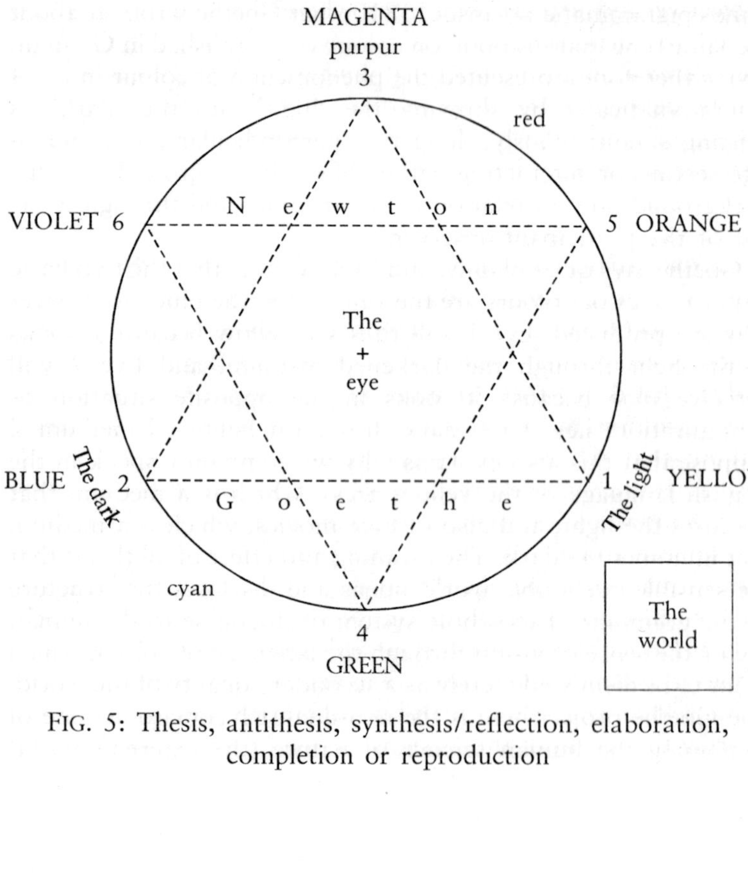

FIG. 5: Thesis, antithesis, synthesis/reflection, elaboration, completion or reproduction

he calls *purpur*, pure red or magenta, already produced in the experiment of the Goethe spectrum, so as to complete the circumference of a circle of colours or a ring of hues; and he encloses the whole phenomenon of colour, now assessed qualitatively as well as quantitatively, around its required natural percipient, the human eye (fig. 5). The student of colour should see with his eye and give full value in his mind to the fact that, in quantitative as well as qualitative terms, there exists a logical and empirical space for this magenta hue, if only *in absentia* in the Newton spectrum, somewhere between the violet or the Italian red which remains on the blue side (left) and the spectral orange-red or the French red which remains on the yellow side (right) of the chromatic circle.

In other words Goethe places at the centre of his scientific concern, not something alone from physics, the purely material and objective, nor yet alone from psychology, the purely mental, subjective and artistic, but from where the two domains are joined (or disjoined), that is to say, the human body when considered from his non-dualist point of view. The human body can be defined in the manner of Goethe as being that part of the external world which mind can know subjectively, and as also simultaneously being that part of the self which one can know objectively. The human senses, far from being merely windows on the world, are the artefacts of mind and labour that mark the frontier and mediate between the facts of the external world and the categories and the values of mind. By the frontier here is meant that point of conjunction and disjunction across which is established the unity in variety of the eye, the mind and the hand or of the faculties and the experience of perception, cognition and volition or of self, the world and the other. I think that the human body was for Goethe therefore the best scientific laboratory, the perfect experimental medium for the interlocution of man and nature, mind and the world, the subject and the object. Fig. 5 thus represents an implicit transformation of the phenomenon of colours from the view of the official Copernicus tradition (Newton) into the view of the radical Paracelsus tradition (Goethe), since taken together as a whole it assumes and proves that man knows the world only in himself as well as that man knows himself only in the world, treating the world of things/signs as one of symptoms, the part and the whole, and our non-dualist knowledge of it to be like the writing and reading of signatures (semiology).

III

THE SCIENCE OF COLOURS

Goethe : Newton
::
Light : Matter
::
Truth : Reality
::
Colour : Rays

Let us now, setting aside the purely polemical part of the debate versus Newton, return to the structure of Goethe's own

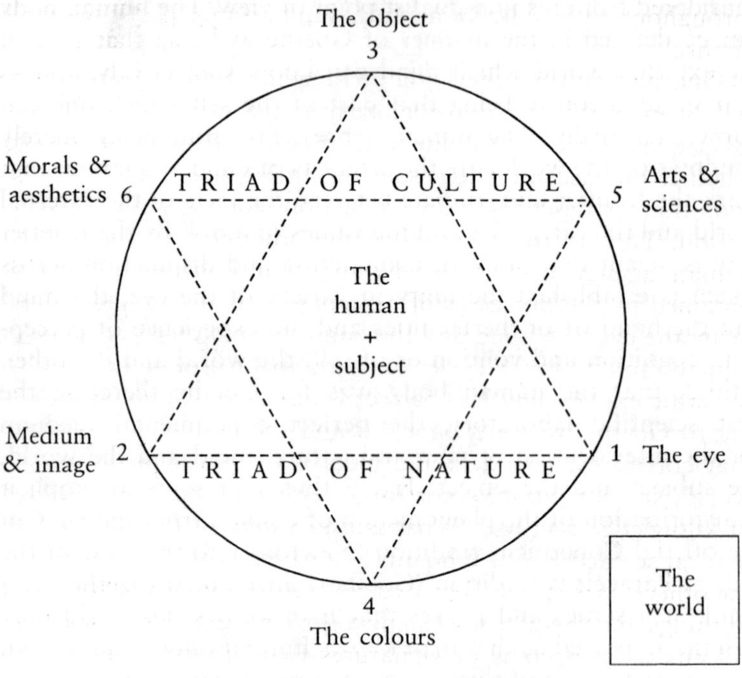

FIG. 6: Goethe's book of colours

exposition in his book, *Theory of colours* (1810) published in German, which is divided into two main parts, each again subdivided into three chapters (fig. 6). The very first chapter is that on the eye of the subject: its focus is physiological: we shall take it up in a moment. The next chapter then moves from the subject towards the object. It is on physics, dealing specially with the (temporary) image formed in the medium and seen by the human eye, the unity of the percept and the concept. The third chapter is finally on the chemistry of colours, i.e. as pigments permanently fixed in the object of nature, science, technique or art. In a Newtonian scheme of course the exposition, even if it were the same in its contents, would be entirely differently arranged in its form, and would therefore possess a different meaning and effect. The successive chapters would follow one another strictly in order of the complexity of phenomena, so that the assumed 'fundamental' science of physics would come first, the chemistry would follow second and the physiology third and last, with the human observer held throughout to be ideally neutral, standardized and out of the picture.

One has only to recall the arrangement of modern school syllabi and textbooks to recognize fully this official developmental sequence of matter, life and mind; and it is maintained whether one ascends the ladder by 'emergence' for the sake of evolution or descends it by 'reduction' for the sake of explanation. The Goethe scheme, on the other hand, puts first the human agency of knowledge, and that knowledge is to be obtained by the processes of perception, cognition and volition or by the eye, the mind and the hand, forming together a single human and natural cycle of participant-observation in the scientific experiment. It is perfectly logical and sensible, according to Goethe, to study first self and the body, the instrument that correlates, separates and mediates between mind and the world, so that we may not forget either of the two questions of knowledge and life, 'What is the system of the world?' (the natural sciences) and 'Who am I?' or better 'Who are we, I and the other?' (art, psychology and religion), and so keep the nature of things and the nature of signs together, e.g. in effect regarding colours as symptoms or signatures that are both natural and cultural at once.

In his very first chapter on human physiology and the eye, therefore, Goethe develops the concept of the after-image of colour by 'reflection' in the eye. Such after-images are known to occur in the eye according to the three diametrically opposed pairs of colours, the three elementary or simple hues (1, 2, 3) invoking the three complex hues (6, 5, 4) and vice versa, on what he calls the chromatic circle (fig. 5). If one looks fixedly at the magenta, purpur or pure red for some time and then shuts one's eyes, one will not perceive the same hue now fading when the stimulus is removed, but one will see a complementary green after-image which of course then fades, and the same applies vice versa: one sees a magenta after-image after fixedly looking at green. The same experimental rule applies between the opposed pair of yellow (elementary) and violet (complex), and between the opposed pair of cyan blue (elementary) and orange-red (complex), which likewise reciprocally invoke one another

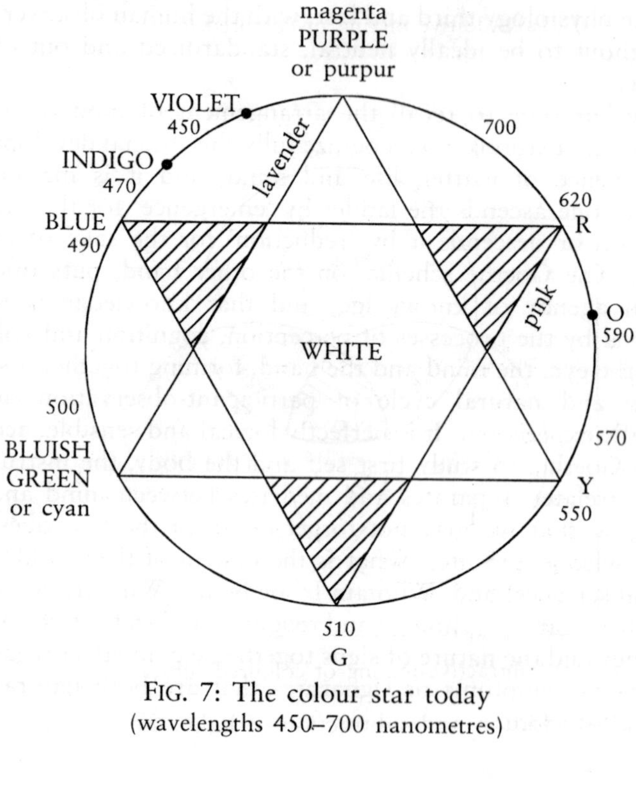

FIG. 7: The colour star today
(wavelengths 450–700 nanometres)

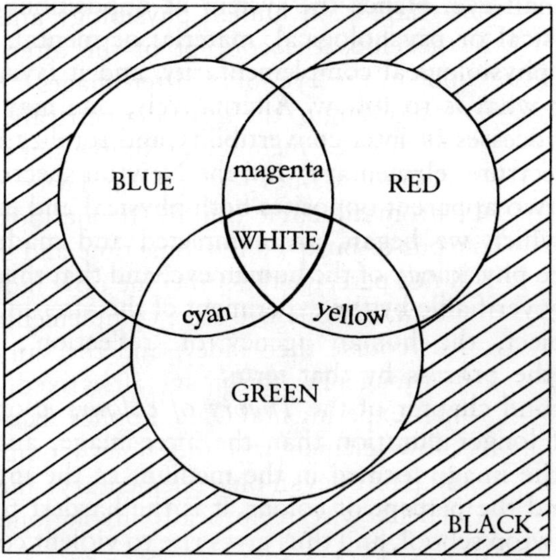

a) Additive mixing of coloured lights

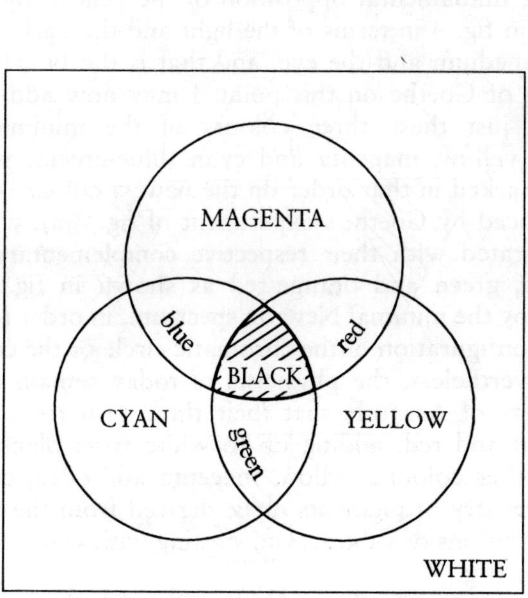

b) Subtractive mixing of coloured substances

FIG. 8: Addition and subtraction of lights and pigments

in the human eye. Hence the system of oppositions is not so much physical or psychological, material or mental, as it is a system of physiological complementarity, and it lays the foundation for what is to follow. Alternatively, one may conclude that the processes of inter-convertibility and totalization of the Goethe spectrum (elementary) and the Newton spectrum (complex), the two apparent opposites both physical and experimental with which we began, are completed and made possible through the physiology of the human eye, and that this is readily empirically verifiable by the experiment of the after-image in the living subject, the human agency of 'reflection', if I may designate the process by that term.

The second chapter of the *Theory of colours* is on physical colours of longer duration than the after-image, and with its focus on the image formed in the medium as the unity of the concept and the percept of colour. It is the hardest to swallow for the Newton school, and thus gives rise to violent controversy when it is considered at all. I have already tried to explain its underlying fundamental opposition of the yellow and the blue as shown in fig. 4 in terms of the light and the dark, warp and weft, the medium and the eye, and that is the best I could do in defence of Goethe on this point. I may now add, however, that it is just these three colours of the minimal Goethe spectrum, yellow, magenta and cyan (blue-green), which one may see marked in that order on the newest colour computers, first produced by Goethe's experiment of fig. 3(a), which were then integrated with their respective complementary colours, violet-blue, green and orange-red as shown in fig. 3(b) and produced by the minimal Newton spectrum, in order to produce the total configuration of the chromatic circle or the colour star, fig. 5. Nevertheless, the physicists of today remain convinced as followers of Aristotle that their three primary colours are blue, green and red, adding up to white from blackness, and that the other colours, yellow, magenta and cyan, can be left to the chemistry of pigments or be derived from the former by suitable additions or subtraction, starting with white and ending with black (fig. 7).

The meaning and effect of this new wickedness are represented in fig. 8(a), the additive mixing of coloured lights (physics), and fig. 8(b), the subtractive mixing of coloured substances (chemistry), which together would yield the colour star/circle of today's school textbook as shown in fig. 7 (which follows the Commission Internationale de l'Eclairage, 1931). The best that can be claimed for this legerdemain is that, on the basis of new and more accurately fine determinations of wavelength/frequency and the reactions of three types of rods and cones in the living eye, it has drawn the yellow, orange and red a little closer together (right), and also pushed the green and the blue further apart to make room for the blue-green or cyan hue in between (left), but leaving the green and the magenta, purple or *purpur* as two paired polar opposites as before. What is to be noticed and regretted, however, is that the centre of the circle is no longer occupied by the eye of the subject, who is out of the picture and is now replaced by the white—but without its percipient so that the after-images are no longer to be seen or considered, i.e. physiology or psychology is expelled from physics or chemistry again, the science of colour disjoins self from the world, and we are back to Newton. Of course in today's nuclear physics the three basic *quarks* or sub-elementary particles come in three colours according to their respective quantum properties or qualities, red, yellow and blue (three primary pigments), but it will be said that this is purely notional (subjective) and it might just as well be red, green and blue (three primary lights).

The frequent mention above of magenta, or Coronation red as it was named in honour of the official robe of this colour worn by Elizabeth II in London, 1953, brings us to Goethe's third chapter, which needs to be read with some knowledge of early modern alchemy in Europe since this is the third dimension that Goethe brings to his unity and struggle of opposites, the chemistry of colour as fixed permanently in the object. The yellow and the blue, when these two basics are combined somehow 'raw', scalar as it were, will produce green, the symbol of earthly peace and harmony (fig. 5, bottom). But if the two are first separately intensified, augmented, developed (*Steigerung*) or somehow 'cooked' (vector), each of them will experimentally approach magenta or pure red, the single hue of their true heavenly advent as well as reconciliation (fig. 5, top). This colour to which Goethe gave the name *purpur*, physically definable only by its physiological opposite green, is the one by which he rejoined the two extreme ends of the unilinear Newton spectrum, the violet and the orange-red, whose reddish quality in common is given no value by the Newton scheme because of the latter's purely mensurational terms of the frequency and/or wavelength of rays. *Purpur*, magenta or pure red, as Goethe said, somehow includes, either in actuality or potentiality, all the other colours, but it is not to be found at all in the Newton spectrum or for that matter in the much-admired rainbow of nature. In early modern alchemy, e.g. Paracelsus, the white when it is put into the furnace passes through the yellow and orange stages or phases to emerge as the majestic red, master of all colours, 'the oriental king' or the rising sun (I suppose). It is of course the colour of the passion of Christ as Saint Basil said when he allocated the four colours, the colour of the blood of the lamb, the mediator, who is begotten before the dawn of creation and before all ages of history and slain from the foundation of the world; while the rainbow of nature (the virgin's girdle) is correspondingly in early modern Europe after the meditations of Birgitta of Sweden the symbol of the virgin Mary, the feminine mediatrix, whose loving intercession with God men may expect at the apocalypse and the end of the world.

From our semiological point of view, the thing to notice is that the triadic form of the minimal Goethe spectrum represented by the upright triangle in fig. 5 is made up of 'elementary' colours here numbered by me 1, 2 and 3, the two polar opposites of yellow and blue resolved or reconciled into a new or original third: thesis, antithesis and synthesis, let us say. Then occurs that vital new process, the jump or the leap from colour 3 to colour 4, which I can only describe for the present as the process of reflection, the 'pure red' being changed through the eye into green, its after-image. After this moment of reflection or mutation in the cycle of transformations, the colour is further elaborated into two more 'complex' colours, colour 5 (orange) on the one side towards yellow, and colour 6 (violet) on the other side towards blue, which completes the triadic form of the minimal Newton spectrum as represented by the inverted triangle of colours numbered 4, 5 and 6. Thesis, antithesis and synthesis are to be necessarily followed, complemented and completed by the processes of reflection, elaboration and reproduction or completion, with the human agency, which is here the eye, occupying the centre of the circumference of forms, the colours of the external world that lies all around. The elementary triad of the Goethe spectrum and the complex triad of the Newton spectrum, shown separately on fig. 3, thus come to be interwoven on the fabric analogy by metaphor and metonymy as alternating neighbours on the circumference of the chromatic circle, and at the same time they preserve their respective physiological oppositions as after-images through the eye in the centre, so that the numerals allocated to any complementary pair of colours together add up to seven, as one can see from fig. 5.

The concluding suggestion that I want to make here is that the relation between the two reflected as well as interwoven triangles of the hexagram or the colour star, the Goethe triad (1, 2, 3) and the Newton triad (4, 5, 6), in some way ultimately represents the relation between the macrocosm (the elementary) and the microcosm (the complex) in the completed Hermetic theory of man and nature, that is, if you will allow as scientific a theory of nature that will include human nature, its physics, chemistry, physiology and psychology. This relation is not altogether or simply that of 'correspondence' as it is termed in the relevant cosmological literature where it is assimilated to the theme of similitude, but rather it is the relation of what I have preferred to call reflection, which is made up equally of similitude and difference so that the microcosm is not simply/only the macrocosm writ small, as it were, but the macrocosm somehow reflected, elaborated and reproduced or completed through the subject or its equivalent. This process of reflection, which I assume to occur in nature as often as in culture, evidently does, among other things, abolish what is the important distinction of metaphor and metonymy. For in the topology of the chromatic circle, every alternating pair of neighbouring simple colours is itself 'intercalated' by a complex colour, so to say, and every alternating pair of complex colours has correspondingly intercalated between them a simple elementary colour. The triadic forms and varied dynamic transformations of our two experiments thus continue to resemble one another in similitude, yet now with the recognized difference that each of them is also the perfect inverse of the other and that the two together represent processes which are formed around the eye of the subject and also objectively in the external world existing all around the circumference of forms: thesis, antithesis and synthesis (the elementary forms) are followed by reflection, elaboration and completion or reproduction (the complex forms).

The second half of Goethe's book of colours is also subdivided into three chapters, like the first half, but there is a moment of change between the two parts which is indicated in a quasi-occult way by the numbering of the sections, which stops short at this point, and it is also indicated by the writing. At any event the fourth chapter explains Goethe's general theory of the phenomenon of colours insofar as this is explained at all in the book. In the fifth chapter he 'elaborates' on the phenomenon through knowledge of the sciences and the arts, including of course the crafts, e.g. painting and dyeing. In the sixth chapter he concludes with the further significance of colour in culture, religion and psychology or morals and aesthetics. At first sight one is rather puzzled as to why the author should explain the general theory of colours only when he comes to the fourth chapter. The rationalist philosopher would have surely put it logically at the beginning of the book. The pure empiricist on the other hand would have arrived at the general theory, apparently solely by the route of experiment and induction, at the end of the book. But Goethe has placed it squarely in the middle of his project at the moment of reflection, as we have defined it, when he changes gear and moves or leaps from the elementary forms to the complex forms, the macrocosm to the microcosm and the sciences of nature to the sciences of culture (fig. 6).

His whole arrangement of the phenomenon of colours as well as the sciences and the arts around the central human subject forms a challenge and an alternative to the official modern classification of primary and secondary qualities, etc. Goethe proceeds here again by the method of the unity and the struggle of opposites setting up polarities, types of opposition and mediation or by the method of thesis, antithesis and synthesis to be followed by the moments of reflection, elaboration and reproduction or completion as I have summarized in fig. 6. Finally, the penultimate paragraph of the book contains a thinly veiled reference to his religion of Jesus and Mary, the mediator and the mediatrix, which I hope to take up on some other occasion for discussion. This was wholly opposed, whether Goethe knew and intended it or not, to the Christianity of Newton, who would have nothing to do with Mariology as is well known from the published sources, and from whose many unpublished writings on religion we now also know that Christology was his other weakest point.

Without going too far further afield into the great corpus of Goethe's scientific work, I shall simply mention that a similar or homologous structure of observation and classification, description and comparison, analysis and explanation will be evident from Goethe's well-known botanical studies, specially the *Metamorphosis of plants* (1790), if we follow the same method of reading, understanding and explaining his discourse of things and signs, symbols or symptoms. The developmental life-cycle of the simple annual plant is decomposed by Goethe into six stages, which begin with the seed splitting into the pair of its two opposed parts, the one going up to the light and the element air (the stem), while the other goes down to the dark and the element earth (the root). The first part then splits further into two, the male (giving) and the female (receiving), together forming the vertical stem and the spiral leaves, the two reunited at a higher level in the flower. So the pair of opposites produces finally the fruit and achieves its completion and/or reproduction through the seed or the bud or through both of them. The simple elementary triad here is that of the growth of the archetypal individual plant composed of the stem and the root, the leaf and the flower, as say children normally draw it shown in a flower pot. Then follows the new moment of reflection of the archetypal individual, who can never alone reproduce himself or herself, into the nature and the conception of the plant as a species, the transformation being achieved through what Goethe calls the life-force at the centre of the life-cycle or the circumference of growth, development and reproduction. This new moment of reflection is then succeeded by 'elaboration' in the fruit and final reproduction through the seed or the bud, which concludes the annual cycle of diachrony. Here again the stages or the phases of growth of the archetypal individual (a, b, c) and the processes of reflection, elaboration and reproduction of the species (d, e, f) are complete polar opposites, each being a kind of inverted mirror image of the other, and yet the two are also perfectly symmetrically interwoven by the temporal sequence of the life-cycle (fig. 9), individuation being followed by holism and the organic whole returning back again to the individual.

Here also then is the unity as well as the struggle of the light and the dark with the world of bodies, the elementary forms and the complex forms, etc., but now with the focus and the rhythms of expansion and contraction, diastolic and systolic, in succession (diachrony) taking the place of the forms of metaphor and metonymy, oppositions and mediations in simultaneity (synchrony), as was the case with the chromatic circle or the colour star (cf. its latest version, but without the concept of reflection, in Boetius, Lauridsen and Lefevre 1998: 38).

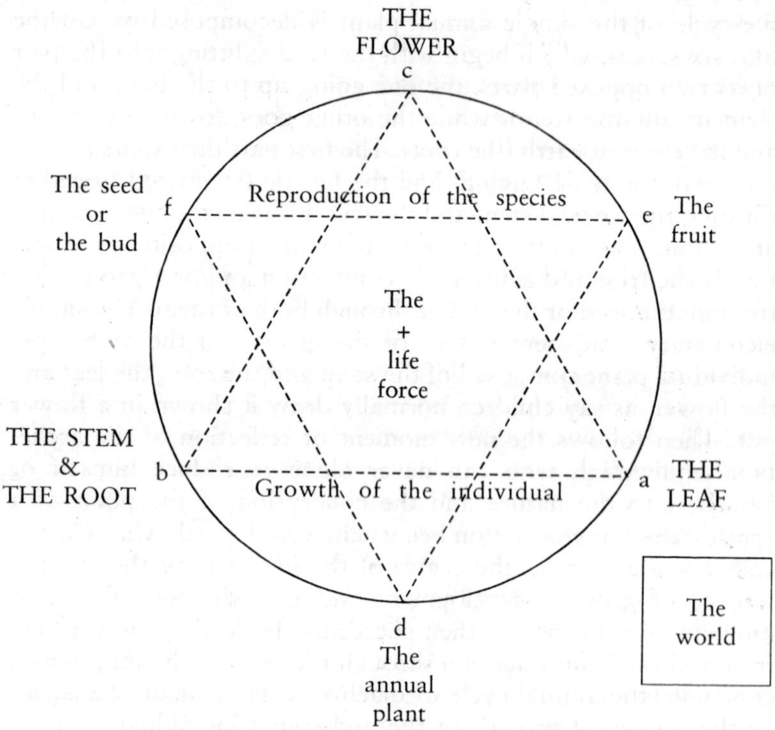

FIG. 9: The metamorphosis of plants

IV

THE RADICAL EUROPEAN UNDERGROUND

> I had separated the heterogeneous Rays from one another... so that there appeared as many Degrees of Colours, as there were sorts of Rays differing in Refrangibility.... For the Rays to speak properly are not coloured. In them there is nothing else than a certain Power and Disposition to stir up a Sensation of this or that Colour... as Sound in a Bell or musical String.... Do not several sorts of Rays make Vibrations of several bignesses, which according to their bignesses excite Sensations of several Colours, much after the manner that the Vibrations of the Air, according to their several bignesses, excite Sensations of several Sounds?
>
> Newton 1931: 122f, 345

After having taken so far a sort of technical guided tour through Goethe's scientific work on the theory of colours in some experimental detail, I should like now to revert to some very general historical and philosophical considerations relating to 'other' modes of thought and codes of conduct, principles of organization other than of the European modernity, in the study of nature. Very often one reads in the established official histories of culture that what chiefly characterized the early modern period in European thought, 1500 to 1650, the first crucial period culminating in the great scientific revolution, was that people went over from a theological or a magical model to a mechanical model of the universe. Only later on in the second or perhaps the third period did they discover or rediscover the organic properties of the cosmos as a whole; and still later in modernity or perhaps post-modernity are the Europeans seeking and finding its spiritual quality or consciousness for study. I am sceptical of the account and not persuaded that this is how it really happened in history; and to try and advance or investigate an alternative hypothesis I had selected and reformulated the Paracelsus tradition and specially the problematics of Goethe, who has been always unjustly represented as an organicist by his critics as well as by his defenders, and have chosen to present his semiological and experimental physics or optics contra Newton rather than simply present Goethe's biology, marked apparently by holism as well as individuation, which he called physical zoonomy.

I think that the simplest way to put the results obtained so far is in the form that the official scientific tradition of Copernicus, Galileo and Newton was always thoroughly dualist, emphasizing both the separateness as well as the difference between the subject and the object, the thing and the sign, symbol or symptom, the percept and the concept, the light and the dark, matter and spirit. It habitually and compulsorily ascribed to the one a purely physical mode of existence and experience, which belonged to the agency of the external world and the study of which was in principle simple, objective, experimental, quantitative and systematic, while it ascribed to the other at best, when it did not lapse into simple positivist reductionism, a purely psychological or cultural mode of existence and experience, which belonged to the agency of the human mind, individual or collective, to history and society, and the study of which was therefore complex, humanistic, interpretative, qualitative and subjective. During the same period, 1500 to 1650, the radical European underground, on the other hand, without in any way denying in the works of its best representatives the valid opposition of mind and the world, the thing and the sign, matter and spirit, always looked further as its scientific duty for a third integrative, correlational or mediatory solution which would have the characteristics of both opposed terms although being itself neither of them. Thus we have the human body viewed as the mediator between mind and the world, the artefact mediating between the fact and the value, or the reading of symptoms (semiology) as mediating between the literal and the figurative, the thing and the sign, or the forms and figures of topology, ratio and proportion as being both qualitative and quantitative.

This European underground of counter-science, truth and method should also have taken up in its theory and practice the study of material culture or the relations and instrumentalities of labour, as Marx and Engels would have had it, as the point of conjunction/disjunction of man (and woman) and the environment, internal and external, the subject and the object, in the way that Goethe had talked of practical gardening or the techniques of dyeing and painting and other crafts as representing the unity of the eye, the mind and the hand or of perception, cognition and volition, but I have not yet found that example in history and society.

Anyhow the official knowledge of the modern period saw only two logical/empirical possibilities as existing between any two things or events and processes, (a) homogeneity and equality or (b) heterogeneity and inequality, i.e. either the two were similar and together or they were separate as well as different, whether in the human mind or the external world or between the two. But for the non-dualist underground counter-science, truth and method there are also two other logical/empirical possibilities to be considered, i.e. the two things, events or processes might be separate but similar (correspondence, equivalence, homology or competition) or the two might be different but together (complementarity, exchange, division of labour), specially in the relation of the subject and the object. We have attempted to show that this mode of thought and experience, which validly encompasses the whole field of all four possibilities, underlies and can explain the truth of Goethe's science of colour (fig. 5); and I am quite unable to agree that the latter's achievements and potential should be excluded and discarded from our science of nature as being somehow inherently non-mathematical (topological), non-quantitative, non-analytical, holistic, vitalist, organicist or humanist in its methodology. It is simply another classification, method and system of sciences (fig. 6), and no friend of the Platonic cosmology needs to claim anything more or ask the enemy of the 'occult' philosophy to concede anything less.

If the unstated dualist-positivist charge against my kind of defence of Goethe is in particular that the colour or hue of magenta, *purpur* or pure red through which Goethe attempted to reunite the qualitative and the quantitative aspects, for example, is not to be found in the Newton spectrum or in the rainbow or in the rest of official physical nature but only among living plants, animals or humans and the artifice of artists, then Goethe's advocate is better advised to plead guilty in court. Goethe stated explicitly that the natural rainbow or the Newton spectrum is in his judgment 'deficient' in the hue of *purpur*, and one may add just precisely because this colour belongs in his system to that part of the external world where the light seen through the dark medium (fig. 4, Eye 1) is somehow put together with the dark seen through the light medium (fig. 4, Eye 2). If this further configuration of mediation, integration or the conjunction of opposites on top of their disjunction or heterogeneity does not occur in official physical nature in the modern European system, then surely human knowledge, life and culture must be allowed the possibility of completing that dualistic conception and that system of physical nature, as with the experiment of the Goethe spectrum, rather than necessarily remaining confined to the further dualism of either simply assuming and continuing it (scientific materialism, empiricism or naturalism) or simply leaving it alone or merely contradicting it (social or humanist idealism, rationalism, hermeneutics or the philosophy of art for art's sake).

Therefore, let us approach this conclusion of the defence of Goethe with my general historical assessment that the function of light in the cosmology of the radical European underground, i.e. including nature as well as culture (or grace), was in effect threefold; and that there were perhaps in all three candidates for that function. *Lux* in the sixteenth century radical Reformation of Servetus, say, had the same functions as the *Logos* in the gospel of Saint John, which was then newly revived, or as the *Pneuma*, spirit or life-breath of the Neoplatonists. For the radical European underground the phenomenon of light represented the mediation of the opposition, upward-striving spirit versus matter or the principle of gravitation, just as the word represented the mediation of the opposition, thought versus deed, or as the living breath represented the mediation of the opposition, the inside versus the outside. The threefold functions of light, the word or the life-breath were thus the multiplicative (source of magnitude), the communicative (motion or change) and the mediatory (conjunction of opposites). This position is (I suspect) already clearly defined for the phenomenon of light by Grosseteste at Oxford (1175–1253), whom some have wrongly tried to annex as the remote ancestor of Newton, who on the contrary almost always stuck at any rate in public to his attempts to reduce the nature of light (rays) to the locomotion of mechanical points (corpuscles), without the power of multiplicative extension, against all comers of whom he knew, e.g. the wave theory. The same

Neoplatonist cosmological background and the full experimental method were also to be found in the Islamic Sufi world, specially in the work of al-Hasan ibn al-Haitham (AD 965–1039), al-Shirazi (1236–1311) and his pupil al-Farisi (d. 1320), whose Arabic writings, method of approach and experimental results were most probably known directly in Paris and indirectly at Oxford.

The present investigation then should help to explain the apparently small technical detail as to why Goethe would have nothing to do with the view that the phenomenon of light is propagated in straight parallel lines (rays), either approximately in nature, as in the arrival of sunlight on the earth, or exactly and precisely, as nowadays in the laboratory laser. I think that Goethe supposed on the contrary that light travels as a sphere expanding from or contracting to a centre or a point, according to the mathematical inverse square law of propagation just like the law of gravity, so that every single ray or segment of it was shaped like a cone and not like a pencil. Goethe's physics of light was therefore altogether one of participant-observation; it concentrated on the study of the situational, transient or occasional image in the medium, the unity of the subject and the object or of the concept and the percept, produced like the phenomenon of the rainbow according to al-Farisi and al-Shirazi. Goethe took some pains to try and show that the whole of the space illumined by the sun through even the largest window is resolvable into the image of the sun, formed inverted except in its dead centre, plus the size of the opening through which the light entered a room. He stated precisely that if we take a series of square apertures of any size we like for the experiment, we shall find that the image thrown on the opposite surface at a distance of nine feet from the aperture will be on every side about one inch larger than the aperture, which very nearly corresponds with the angle of the apparent diameter of the sun as seen from the earth.

If the two attributes of extension and duration in space and time are defined as the characteristic modes of existence in the external world, and if we forget or hold as neutral for the moment the human subject, then the phenomenon of light and its propagation would be Goethe's model of the external world whereas matter and gravitation would be correspondingly

Newton's model of it. The former is clearly in the tradition of Plato, giving first priority of perception to the sense of vision, while the second is in the tradition of Aristotle, giving priority of experience to the sense of touch above the other senses in our knowledge of the external world. In the modern European victory of the official scientific revolution over the counter-science of the non-dualist underground, the other self of the European modernity, it was Newton who made the study of light and the study of gravitation or of optics and mechanics separate and different forever; and what he had divided no other man, not even Einstein, has been able yet to put together again.

But in order to be as fair as possible and to bring the discourse and dialogue up to date or at any rate to the mid-twentieth century, one must mention the standpoints of both Einstein and Heisenberg, the latter of whom has directly expressed himself on the Newton versus Goethe issue. Firstly Einstein, as he said, was not a revolutionary nor a radical of discontinuity in modern physics, and had for many years wanted to reconcile the material or mechanistic theory with the field theory or the wave theory, but on strictly Newtonian terms or some development of them. It was to be in the same positivist-dualist tradition which holds that all scientific theories are necessarily either models of reality (nature) or only models of our human knowledge of it, models of truth (culture), and that between the two there must be only degrees of sameness, homogeneity or heterogeneity, as if these were the only two possibilities and their duality were sacrosanct to the exclusion of any third or fourth possibility. He insisted that, insofar as the propositions of mathematics referred to reality, they were not true or certain; and so far as they were true or certain, mathematics did not refer to reality. The Newtonian ambition of Einstein had proved to be impossible of realization time and again in both physics and philosophy. I was thus advised by experts to appeal directly instead to Heisenberg, the pillar of quantum mechanics and the uncertainty principle, complementarity, etc. Alas, I discovered only too soon that he had explicitly opined in 1941 that it is the very self-same unity of Goethe's science of colour, which attempted to cut across apparently separate and different systems of study and domains of nature, physics, chemistry, physiology and psychology, that is unacceptable to the modern physicist, meaning himself as well as Newton, and for that very reason of seeing unity in variety.

Our conclusion must be that the radical European underground wanted to study the phenomenon respectively of light, the word or the life-breath as being the correlation, the mediation or the unity in the duality of the thing and the sign, the object and the subject, macrocosm and microcosm, the percept and the concept, expressing their mutual relations of competition as well as cooperation, homology, symmetry and complementarity or the dialectics of equivalence, correspondence and reflection. The official European science of modernity was determined from the beginning up to the end never to recognize as 'scientific' any general cosmological theories of the trans-systems unity of matter, life, mind and spirit or of nature and culture or grace. Modern western science and modern western culture as twin children of the European modernity were bound to allow to exist between them only dualistic theories of the unmediated heterogeneity or the equally unmediated homogeneity of such terms as sets of things or signs, the external world and the human mind, which according to Goethe's showing is an altogether untrue, self-alienated and disastrous Faust's choice of the structure of discourse in the projects of science as well as of culture, and from which Goethe and the radical European underground valiantly tried to rescue all of us in the other tradition of Paracelsus and an alternative non-dualist modernity. Thus instead of the science of matter and motion or energy, we have in Goethe the science of light and colours, instead of the search for mensuration, analytical traits and their fixed relations, we have the archetypal phenomenon (*Urphaenomen*) and its metamorphoses or transformations, and instead of explanation in terms of the laws of cause and effect, constants and invariants, which was forgetting the presupposed initial situation and boundary conditions of the system, we have the part and the whole, macrocosm and microcosm, the subject and the object, and the search for the variety of their relations. The purpose for the self is not to know and then conquer, or alternatively to destroy, the world, but to cosmologically unite with it through active reflection in consciousness, lexis and praxis.

4

The Manhattan Project and the End of Modernity

> Faustus, these books, thy wit, and our experience,
> Shall make all nations to canonize us.
> As Indian Moors obey their Spanish lords,
> So shall the spirits of every element
> Be always serviceable to us three.
>
> Marlowe, *Doctor Faustus*, 1588

> The venture of which I write in this book represents, in the opinion of most engineers, scientists and industrialists interviewed by me, the greatest single achievement of organized human effort in history.
>
> Groueff 1967: 5

I

THE MANHATTAN PROJECT (1945)

The new semiological theory of the European modernity, if indeed it be new, offered here is (a) that, firstly, modern western science and modern western culture were born together as twin children of the new schismatic revolution of Europe which took place between the Renaissance, the Reformation and the counter-Reformation, starting in about AD 1500. (b) The coming of age of the European modernity occurred when its threefold separation of science, religion and politics or of faith, knowledge and action was officially and formally established in the institutions of civil society, the church and the state. This defining event was marked in Britain, say, by the simultaneous separation of the new Christians, the Nonconformists, and the foundation of the Royal Society of London for the propagation of natural knowledge by Charles II in the mid-seventeenth century, 1662. To these two hypotheses I now want to add (c) that the great implicit dynamism of the European modernity in science as well as in politics took altogether (another) three centuries, say, to find its explicit internal and external limits, which it has now passed in the mid-twentieth century. The twin high peaks of its fulfilment were of course the two revolutions of industry in Britain and of democracy in France, c. 1800, although in positivist and dualist style they never joined together to produce what would deserve to be called industrial democracy. (d) That is, fourthly, I think that its internal limit was marked by the political flames of endo-cannibalism that engulfed Europe and Russia under the similar totalitarian and autocratic regimes of Hitler and Stalin; and that its external limit was marked at almost the same time by the scientific-experimental explosion of the first atomic bomb and its immediate military use against Japan (1945). From my non-dualist point of view, then, the construction of Hitler and the destruction of Hiroshima together formed the last great civilizational event of the European modernity and its military-industrial-scientific complex.

The expression of scientific as well as political cannibalism, which I use advisedly, refers in the first place to the scientific and racist torture chambers of man at Auschwitz, Dachau and Buchenwald, which had selected and slaughtered six million fellow men, women and children by 1945, the end of the war. (The phrase 'industrial extermination' was used by the sociologist Raymond Aron at the UNESCO symposium on social development, 1960.) The infusion of Nazism and Fascism at home in one form and another was one side of the modern industrial capitalist coin. Capitalist European imperialism abroad, also driven conjointly by industrial science and political racism in modern times, was the other side of the same coin. The ultimate military weapon which was manufactured by the positivist laboratory and bureaucratic organization of Los Alamos, USA, was put into practice without warning against non-combatant non-western outsiders, and so far notably only against the latter, within a mere three weeks of its invention.

Science, industry, bureaucracy and politics under the modernist regime had achieved separately and together the ultimate therefore in endo-cannibalism as well as exo-cannibalism by the mid-twentieth century. In relation to both of these aspects, the internal and the external, the hypothesis is that science and politics, the truth and the power or philosophy and technique, had never before been apparently so close outwardly but so far apart inwardly as in that total western climax of 1945. The institutions of civilization were at one and the same time inwardly divided against themselves but externally united against the non-elite and the non-expert, the third party, both national and international. What had been at first sight simply oppositions and contradictions to be perhaps exploited, resolved, mediated or transcended had been turned by infinite regress into the nightmare of pure dissociationism, self-deception and self-alienation by the end. We shall first interpret this diagnosis below and then come to some prognostic conclusions from a post-modern or a third world and non-western point of view.

Let us recall first the threefold disposition of science, the military and politics in the modern west as illustrated by the so-called Manhattan engineer district, which was set up in 1942 to help shorten the Second World War. To deliver the atomic bomb within three years the Manhattan project secretly expended $US 2,000,000,000 and employed more than 150,000 scientists, designers and technicians. The commanding officer of the organization, Lieutenant-General L. R. Groves of the United States army, said that it was one of the greatest combined scientific, technical and industrial achievements in history; he should have added also of culture, politics and economics. It had been the conjoint responsibility of the US war department and the office of scientific research and development, i.e. after its political clearance.

Already, Jewish and other refugee scientists from Germany had transplanted the ideas of nuclear fission and the atomic bomb to America, where the necessary research was first undertaken at the three universities of Columbia, Chicago and

California. The united effort was shifted into high gear a few weeks after the United States entered the Second World War on the side of the Allies versus the Axis powers. In 1943, a highly secret scientific laboratory of the Manhattan project was set up on a remote isolated mesa situated at Los Alamos, New Mexico. The site was chosen by Oppenheimer, the brilliant and sensitive physicist who later showed contrition when it was both safe and useless. He was pulled out of California and quickly put in position as the director of what was claimed to be 'the best-equipped physics laboratory in the world'. The project was all set to make a two-pronged attack. The nuclear materials required were being produced elsewhere either by the transmutation of uranium into plutonium, which was a new element and not an isotope of uranium (Hanford, Washington), or by the separation of pure uranium 235 by the electromagnetic process and/or the gas diffusion process, whichever proved effective (Oak Ridge, Tennessee). At the same time and without waiting for the lethal fissile material Oppenheimer and his able scientific team worked devotedly on the atomic bomb, its design, basic physical research and final assembly and delivery. Within close on two years they had managed to produce three atomic bombs of two different types, only the first of which of the plutonium 239 type was pre-tested in the US, while the other two specimens were soon air-dropped on Japan (see fig. 10).

The very first ball of atomic fire on earth produced by the secret test at Alamogordo that was obscenely code-named Trinity arose at 0530 hours on 16 July 1945 out of the desert of New Mexico to a height of one mile into the sky. It was seen to be 'brighter than a thousand suns'. It apparently reminded Oppenheimer at the time of that phrase from the Bhagavad-gita, where it was used to describe or to try and describe the effulgence of God. Oppenheimer's ball of unearthly light, however, was immediately followed by a mushroom-shaped radioactive opaque cloud which arose to a further height of five to seven miles into the sky. The day of the atom had dawned, or perhaps the laboratory of truth was turning into the factory of power, as always under the modern west. The telegram on plain paper placed before Churchill, then already on his victorious tour of Berlin and Potsdam, said it in just three words: 'Babies satisfactorily born.'

The Atom Age Is 15 Years Old: A Look Back Reveals a Secret

By WILLIAM L. LAURENCE

Saturday marked the fifteenth anniversary of the detonation of the first atomic bomb on the desert near Alamogordo, N. M. It was at the time the war's topmost secret.

As those present watched the mushroom-topped radio-active mountain light up the dark desert with the light of many suns, and heard the awesome roar that echoed from the Sierra Oscuro range all around, they knew that they had attended the birth of a new age.

Although the weapon, known by its designers as the "gadget," was the first to be tested, it can now be disclosed that it was actually the second atomic bomb in existence at the time.

The first atomic bomb, known among its makers as the "Lean Boy," had already been made.

Its component parts were ready to be shipped, by the cruiser Indianapolis and by B-29's (the largest wartime bombers) to the small island of Tinian, in the Marianas. This had been chosen to serve as the base from which the atomic bombers were to take off for their chosen targets in Japan—Hiroshima, Kokura, Nagasaki and Niigata.

The two weapons were different both in the fissionable materials they used and in design. The first bomb, used three weeks later over Hiroshima, contained uranium 235 as its fissionable material. The material was enclosed in a long gun barrel in two parts, at opposite ends of the barrel.

The two parts together made up what is known as the critical mass, an amount that would explode automatically once the two subcritical masses were brought together at the desired moment. This was accomplished by a proximity fuse when a gun mechanism at one end of the barrel fired the smaller of the two parts into the larger part at the other end of the gun barrel.

The second atomic bomb, the type tested on the morning of July 16, 1945, at Alamogordo was known among its makers as the "Fat Boy." Its fissionable material consisted of the man-made element, plutonium, element 94, the existence of which was at the time a super-top secret.

It was designed to be exploded by an entirely new principle, never tested before on a large scale, known as implosion, a word meaning a bursting inward, as contrasted with an explosion, which means a bursting outward.

An implosion is a way to bring about a critical mass by increasing the density of the fissionable material by the application of sudden and equal pressure at all points. For example, a given mass of material may be subcritical at normal density, but by the application of great pressure, evenly distributed on all sides of the mass, the density of the material is increased from subcritical to critical. The reason is that in the denser material more atom-splitting neutrons are kept from escaping outside the mass.

It was virtually certain that the "Lean Boy" type would work, but since the "Fat Boy" used the entirely untried principle of implosion, as well as a new fissionable material, it had to be tested before it could be used.

When the test turned out to be even more successful than the most optimistic of its designers had dared hope it was realized at once that the "Lean Boy" type, which required the use of much larger quantities of fissionable material, was obsolete even before it was used, but there was no time to redesign it. As a result, it was the only atomic bomb of its kind ever made or used. It was the bomb that devastated Hiroshima.

It was the "Fat Boy" that destroyed Nagasaki.

The success of the implosion principle also brought about the first experimental proof that a hydrogen bomb also could be made.

FIG. 10: The atomic bomb is fifteen years old

Source: New York Times, *18 July 1960*

For the benefit of today's young people who may not remember all the details, I must recall that the military position at the time was that Germany was already thoroughly defeated and had surrendered and Hitler was known to be dead; that Mussolini had been also overthrown in Italy two years earlier and was now dead, summarily executed by partisans of the Allies; and that imperial Japan was tottering on its last legs and about to fall, according to Churchill's own testimony.

Nevertheless, under the modernist regime the other two bombs of the Los Alamos physicists had to be delivered. What was supposedly made for Germany was delivered and in the best Nazi way to civilian Japan. The uranium 235 bomb was thus experimentally tested for the first time in typical vivisectionist style by dropping it on Hiroshima at 0815 on 6 August 1945 (Japanese time), when the maximum number of persons would suffer the maximum exposure by being either on their way to work or outside for drill, not to mention the 10,000 children engaged in making fire-breaks. The already pre-tested plutonium 239 type was also dropped on Nagasaki three days later, again without warning, at 1102 on 9 August 1945 (Japanese time). Each type of bomb was approximately of the equivalent power of 20,000 tons of TNT, although this figure was subsequently differentiated for the two sub-types. Nowadays, at the present and recent rates of progress, the reckoning would be more than a thousand-fold, i.e. in terms of megatons (× 1,000,000) rather than in kilotons (× 1000) of TNT. Some two thirds of the city of Hiroshima was instantly and completely destroyed by all accounts. Of its estimated less than 300,000 inhabitants some 70,000 or more were killed within a few seconds, while another 70,000 or so were wounded, many of them in new and strange ways to die horribly and incurably later. The story of Nagasaki, the sister city of scientific death, was precisely the same: a quarter or 40,000 dead at the time; the like number injured; and the city left afterwards only 'a graveyard with not a tomb standing'. The toll of the dead and the injured mounted eventually to a total of 300,000 souls for the twin bombs, one of the plutonium type and one of the uranium type; both of them were as predicted very high explosives, super-incendiary and spread long-lasting radioactive poison.

The combination of science, the military, bureaucracy and politics in the Manhattan project was completely successful objectively, therefore, but its subjective effect was little noticed at the time. It distributed the responsibility in such a way that no one should own it: the subjective moral and ethical questions of power disappeared without trace into the interstices between the physicist (Oppenheimer), the general (Groves) and the President of the United States. The alliance between science, industry, labour and politics, judged to be rationally expedient for the sake of one or the other good cause of progress or project of innovation from time to time, was always only a temporary marriage of convenience. The dualistic separation of a value-neutral technical science and organization from all questions of politics or philosophy, ethics and ideology was, on the other hand, fundamental, necessary and permanent. The two aspects of this dualistic regime of civilization had divorced long ago by mutual consent to form the modern world, and they were only ever to be seen together for appearance's sake or for the sake of the children, so to say, who might turn out to be either good or bad on their own.

II

THE SCIENCE OF VIVISECTION

> A little more than a year ago we first met Professor [Masao] Tsuzuki, the Tokyo radiologist. He came to the bustling temporary headquarters of General MacArthur in that first confused week of the occupation to tell us about Hiroshima. He is a small, quiet man of great dignity and used to deference. He showed us his reports of the radiation casualties in Hiroshima, and with an air of detached, scientific enthusiasm, he gave us reprints describing his work as a young man in Philadelphia. At that time he had induced radiation deaths in rats by acute doses of X-ray. The symptoms he had noted so carefully in the rats who died twenty years ago were plainly parallel to those familiar by now in the patients who crowded the temporary hospitals of Hiroshima. In real concern Professor Tsuzuki [who later died of cancer, 1961] grasped my arm and said: 'I did the experiment years ago, but only on a few rats. But you Americans—you are wonderful. You have made the human experiment!'
>
> Morrison 1946: 5

> The components of Little Boy—gun, projectile and target—were ready. If the theories were correct, if the tests were conclusive, the bomb should now explode. But unlike the plutonium programme, no test was planned for the complete uranium bomb—there simply was not enough material for two bombs. Production of uranium 235 was slow at Oak Ridge and the very first weapon that could be assembled was therefore earmarked for combat release. If Little Boy was to be used in August, no prior test or demonstration was possible.
>
> Groueff 1967: 380

UNCLEAR ON NUCLEAR

> ...The Hiroshima bomb was not tested at all and was totally different from the 'trinity' device and the Nagasaki bomb.
>
> *Times of India*, New Delhi, editorial, 15–16 October 1997

I now want to focus special attention on the fact, which was carefully concealed at the time and only began to emerge partially at the Rosenberg spy trials several years later, that there were made by the Manhattan project altogether only three nuclear weapons, specimens or devices of two very *different* kinds. The one used at Hiroshima was a 'gun type' uranium 235 device, while the one used at Nagasaki was of the smarter 'implosion plutonium type', and General Groves only confirmed in 1962, when the Manhattan project files still remained inaccessible, that the two bombs used were the only ones ready at the time. The plutonium 239 kind was thus readied in two specimens, the prototype as we may call it duly tested at Alamogordo, after which date the responsibility passed from the scientists and engineers to the military and the statesmen, who among them decided to drop the second specimen on Nagasaki. We are shown here, therefore, the two spheres of respective special responsibility clearly subdivided and meant to operate in a sequence of time (fig. 11, the first set, a, b, c). But there was made by Los Alamos and ready only one specimen, bomb or device of the uranium 235 kind. So when this latter was dropped *first* upon Hiroshima, one can be sure that the scientific and technical, military and political yields were all together aspects of a single action and decision of civilization and the nuclear science project.

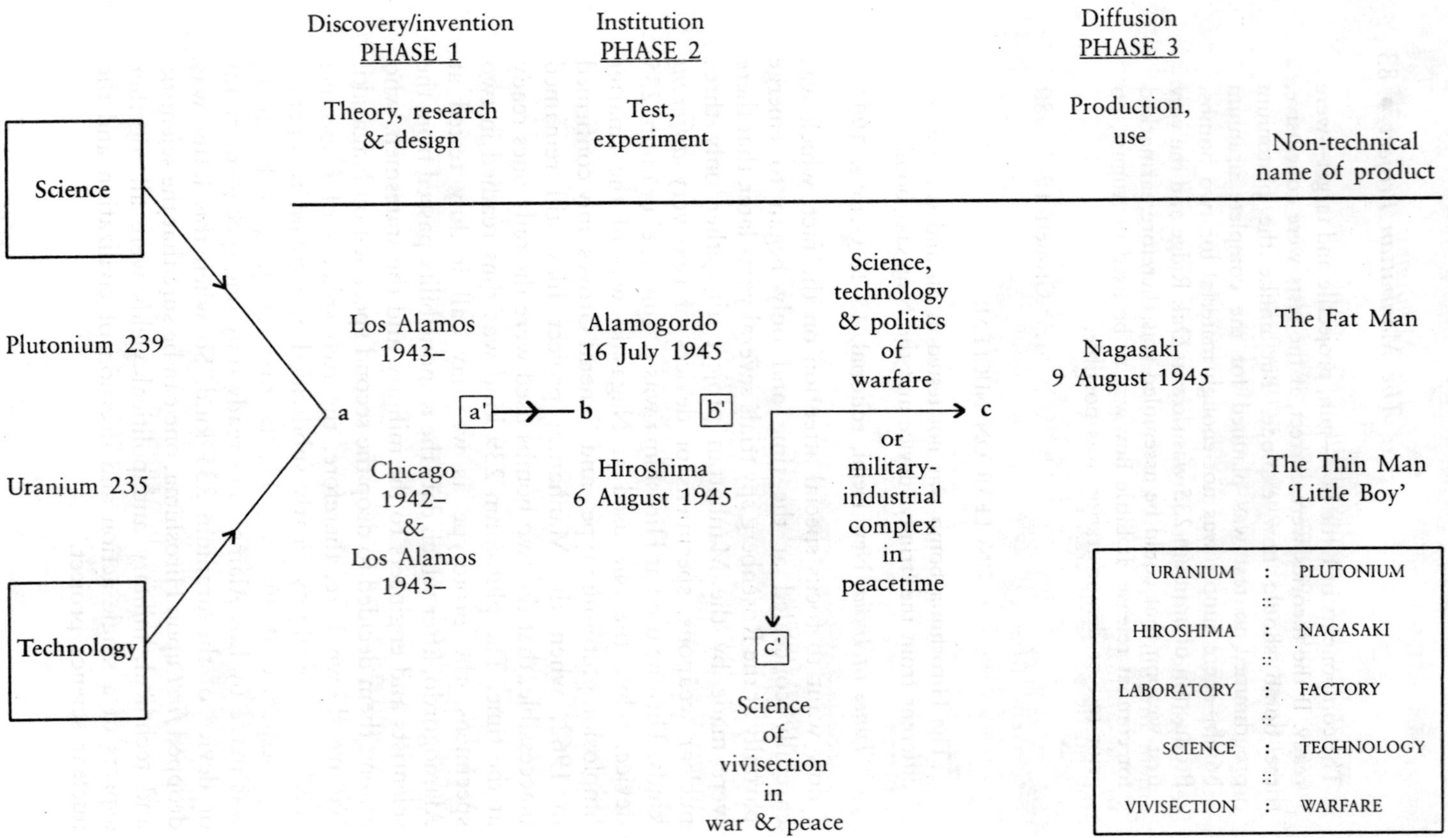

FIG. 11: The Manhattan project

What had been achieved for man or the human species, as we are proudly informed by the monument erected later by the authorities, on one winter afternoon inside a pile of graphite on a disused squash court of the University of Chicago at 3.25 PM on 2 December 1942, namely, the first experimental self-sustained and controlled nuclear chain reaction in the element uranium, was quickly arrested after half an hour's trial by its authors. It was subsequently let out as the genie of the bottle to run its full course, without acknowledgment after Alamogordo, as both a laboratory experiment and a weapon test over Hiroshima, where it was required not in order to win the war against Japan but to complete the scientific theory of a critical mass and nuclear explosions, making it a victory for science over politics (fig. 11, the second set, a', b', c'). Here one can see naked, therefore, the underlying *science of vivisection* of Faust's modernity whereby the end, whether of knowledge or of power or both, justifies the means; and this will also help to explain incidentally why the effects and the victims of Hiroshima (U 235) and of Nagasaki (Pu 239) were kept so carefully separated in all follow-up medical and other studies made in Japan by the Americans. But alas, by the same unfortunate dialectics of dualism in relation to faith, knowledge and action we might just as well arrive at the opposite conclusion, technicism versus expediency, where the means would justify the end, since the two are completely dissociated in this civilization.

The science of vivisection will stop at nothing, in other words, that it might find smelling 'technically sweet' in the phrase of Oppenheimer, and 'what is technically sweet will be produced' whether it should simultaneously mean 'knowing sin' or getting 'blood on one's hands', also the US physicist's phrases, which he would never be able to put together in his time (technicism, the means justifies the end as desirable if unnecessary). On the other side of the 'iron curtain' too, when it was raised, we find the so-called 'peacenik' of progress, coexistence and freedom Andrei Sakharov, youngest member of the Academy of Sciences, who had played a leading role in the development of the Soviet thermonuclear bomb perfected in 1950 at almost the same time as the first US H-bomb, writing in his memoirs that the physics of atomic and nuclear explosions was a 'genuine theoretician's paradise' (1990).

The Allied statesmen all tell us with one clear voice that it would have been worse without the bomb (expediency, the end justifies the means as necessary if undesirable), and who can prove them wrong? Truman, who finally ordered the delivery of the babies, says that his advisers had told him that an estimated 500,000 American soldiers would be lost in the invasion of Japan. The evidence of Churchill is somewhat more ambiguous. On the one hand the great British statesman thought that the bomb was a godsend to save the lives of friend and foe alike. On the other hand he writes on reflection that

It would be a mistake to suppose that the fate of Japan was settled by the atomic bomb. Her defeat was certain before the first bomb fell, and was brought about by overwhelming maritime power (Churchill 1954: VI, 559).

James Olson reviewing Herbert Feis and referring to the official US strategic bombing survey of 1947 concluded that certainly before 31 December 1945 and in all probability before 1 November 1945, the invasion date, Japan would have surrendered even if the twin atomic bombs had not been dropped, even if Russia had not entered the Pacific war and even if no US invasion had been planned or contemplated (*Bulletin of the Atomic Scientists*, 1962, 17[2]: 35). It is also well known from several sources that the Japanese had been already trying for months, through Russia and perhaps the Swiss, to surrender, accepting all the US terms with but one reservation: they insisted on keeping the emperor.

Anyhow, by the time that the Manhattan project was wound up in 1947 and its responsibilities turned over to the US atomic energy commission, other military-industrial-scientific organizations and projects just like it had already sprung up under other great powers by competition or cooperation or both. The USA itself was the first power to go on to prepare still more potent horrors for the non-elite—and Oppenheimer always insisted under official investigation that he had not lagged behind in the effort. The 'categorical imperative' of modern western civilization and critical rationality had long since evaporated from science as well as politics and left behind only naked technicism and/or expediency to rule instead. From one such brilliant project of progress to the next, the statesman or the scientist and the man, or the 'system' and its meaning and consequences, intended and unintended, or in Durkheim's terms its collective consciousness and its collective conscience, were completely alienated from each other. But all this was surely both necessary and desired as being part and parcel of the original Faustian contract of the division of labour between the truth and the power, fact and value, theory and practice, the ends and the means, which had finally disclosed itself to be essentially the application of a false, evil, ugly and useless logic. At the same time, modern positivist dualism in relation to faith, knowledge and action had finally shown itself disintegrated into a kind of double monism of two separate and different worlds, i.e. modern western man had progressed from perversity to a kind of insanity, and the system of positivist dualism, established in 1500 and grown to flourish, 1650–1800, had degenerated by the mid-twentieth century into godless, self-deceived and endless devilry.

All later self-disclaimers apart, the Manhattan project that Oppenheimer finished along with Truman in 1945, Einstein himself had avowedly begun with Roosevelt in 1939. The greatest physicist since Newton stated afterwards that he had at the earlier date signed his famous confidential letter to the President of the USA and so opened the road to Los Alamos chiefly out of his fear that the Nazis might get there first. The letter was urgently drafted and brought to him by another Nazi refugee and physicist, Szilard, who also turned pacifist after the bombs had arisen, quite in the fashion that Sakharov and others did, thereby turning their respective roles in the project(s) of means into ends. Perhaps Einstein had once fondly imagined that he had put the human observer back into Newton's purely mechanical principles of nature, at any rate in the general theory of relativity, but this single action, at once political and scientific, was to turn him soon into the most helpless observer of the most inhuman use of his theories, $E = mc^2$, etc. By the occult logic of modern western rationality, the Jew and the pacifist who had only wanted and tried to be rid of Hitler and his wars lived in effect to provide the formula for the horrors of Hiroshima and Nagasaki, and, as if that were not enough, also for the nuclear arms race between the west and the east that greatly accelerated after 1945 in peacetime.

(For the role of Einstein in the nuclear bomb project, which was crucial in 1939 and 1940 and also later, see Clark 1971: ch. 20.)

Therefore, I want to arouse in the mind of the reader a suspicion, if it is not already there, that the men who set out to overcome Nazism by military force were not entirely successful in their venture. In the end, the theory and practice of political, military, bureaucratic and scientific dualism, if not of Nazism itself, had found a much deeper and wider following than before; and what was supposed to be justified to put out the domestic fire of the European modernity had ended by setting the neighbour's house aflame. The whole united effort of science, the military, bureaucracy and politics in effect subdivided its leaders as well as its followers into three equally unfortunate categories. (a) Some people were turned by the dialectic into Nazis themselves: they were supermen of the modernity project who would be gods or at any rate devils. (b) Others, who had less pride and perhaps more love, were rendered helpless or themselves became merely cynical participant-observers. (c) Still others apparently bravely carried on the world's business in two minds, using the one for the end and the other for the means. Between the ends and the means, etc., the shadow that had fallen had lengthened to infinity. Either everyone had neglected to read and be warned by Blake instead of only the Bhagavad-gita, and so failed to recognize the English mystic's Wheel of Fire that traversed from west to east against the current of Creation. Alternatively, one may say that perhaps they saw it all right in the middle of 1945 but heroically postponed all idealistic moralizing, so to say, until after the second coming of Christ. After all in psychological terms schizophrenia was not a sin against God or a crime against man but a perfectly natural, realistic and practical way of life for the self under this regime of unlimited dualism.

We are entirely misled, therefore, into supposing that the monuments of Hiroshima as well as Auschwitz were the separate results merely of some unfortunate fortuitous and contingent links between science, the military bureaucratic organization and politics, which should otherwise be treated as three autonomous spheres of responsibility. On the contrary, I would say that their inward divorce and mutual alienation was arranged by and seen as necessary and desirable for the European modernity in any and every form or field; and that therefore no good was to be expected from that relation or non-relation in history and society by virtue of inherent human necessity. For example, as to the conjunction between the science, technology, medicine and political economy of life and death, this apparent modernist unity of the laboratory and the factory or of discovery and invention in truth itself produced or represented the alienation of the species man from nature and from the self as much as from God and the Christian church. From a purely historical, social and cultural point of view, the only possible valid comparison of Hiroshima was with what Hitler had just finished doing to the Jews at Auschwitz in Poland. The two of them were but two moments, events and monuments of that unique modern way of life and thought which had produced such a lethal, unanswerable and irresponsible mixture of the scientific and the power-racist impulses of man as the world had never seen before. I think that it was with the politics of racism as much as the physics of relativity in mind that the west had set out freely to discover the use of the equation between the mass and the energy of matter in the twentieth century.

No scientist of my acquaintance, however, will accept the blame for what he or she will term the *bad* side of applied science. Like the statesman, but only in reverse, the scientist will argue and shift the responsibility for it on to varieties of nationalism, totalitarianism, imperialism or capitalism, the ruling class or human nature and greed or what not, for example, like the British physicist P. M. S. Blackett who wrote that Hiroshima was meant to be a political lesson and demonstration for the Russians. He or she will claim only the *good* side of applied science to be inherently related to the system of knowledge, and attribute the rest to the workings of political economy. It amounts to supposing that if the French positivist Saint-Simon's gentlemen of the first class, the scientists, did happen to bring happiness to us occasionally then we should be grateful and give the credit to modern western science; but if perchance the 'empirico-mechanico-dogmatic torture chamber' of nature, as Goethe called it, should descend from the same positivist heaven to slaughter the innocent, whether non-combatant Jews, Slavs or Japanese, we should always debit the blame elsewhere.

Everything is alive in dualistic parts and nothing is whole in this civilization any longer: truth and reality, the good and the bad, life and death, knowledge or contemplation and action, the spirit and the form. God is dead to the world of course; but the human species and the science of nature do not look so good either—at any rate to the non-expert and the non-elite: 'God is dead, and I am not feeling too well either!'

Yet it is today all the more desirable for this very reason to weigh up the ends and the means by calling into question the modernist regime as a whole. Statistics of the kind quoted by the Stockholm international peace research institute's annual books show that some 400,000 scientists, engineers and technicians, i.e. roughly half of all the world's trained science manpower, are at present willingly committed to and engaged in new destructive weapons research and development (1975); scientific secrecy and political chicanery obscure more recent and detailed estimates. The resultant atomic, bacteriological and chemical weapons of war are by now well established to have been used freely against non-western civilian non-combatants, and so far against them only, in the conflicts of Japan, Korea and Vietnam respectively. Of its ten million killed, the First World War's dead were 95 per cent soldiers and only five per cent civilian. During the Second World War the proportion was made almost even—of its 50 million killed, 52 per cent were soldiers and 48 per cent civilians. After 1945, by the time of the war in Korea, the proportions were completely reversed for the future which is ours: of its nine million killed, only 16 per cent were soldiers and 84 per cent were civilians: this is the net result of making 'not heroism but technology' the winner of war. This is more or less the world situation today, I am afraid to say, after the fall of both Hitler and Stalin, whose evil still lives and spreads after them.

I will now try and sum up in three short points the prognosis and the conclusions that seem to me to follow from our theory of science, politics and culture under the modern western regime.

Firstly, we, the non-elite and the non-expert, should resist the usual intellectual error that would have us find separate and different causes for the good and the bad sides of the European modernity. Both its science and its politics, the successes and the failures, the internal and the external, all have that common quality, the unity or rather the duality that gives modern western civilization its name and form as a mode of existence of the human species. I wished to say that the spirit and the form of the whole modern history which culminated in the twin horrors of Auschwitz and Hiroshima, militant Fascism at home and racist imperialism abroad, depended upon that structure of science and politics, economics or ethics and of philosophy and technique, whose foundations had stood divided by mutual reciprocal consent for several centuries past, from about 1500 to 1945.

Secondly, we must all hope and pray that, its internal and its external limits having been marked and passed in history and society, the day will come soon to fold up the structure of the modern western regime for good and all. When this most prodigious misadventure of the human spirit, so dynamic and yet so God-forsaken in its dualist form, is ended by being completely abjured or renounced from within by repentance and atonement, only then perhaps will the truth and the reality be able to meet again in self and the world, the subject and the object and in their relation. It should be possible then again somehow to speak of truth, since it includes possibility as well as reality, as being the measure of reality and not only as its picture or reflection, and correspondingly to speak also of reality as being the realization of truth.

Meanwhile, it will be an elementary mistake of scientific method to suppose that its object of study is reality when it should be understood to be truth. Truth emerges from the dialectics of what is the case (present) and what is not the case (absent), the world of reality and the still wider world of possibility. Thus and only thus, truth enables us to understand as well as to describe the reality, and even to judge and change it as well as to change oneself. This structure of truth and method in science is what we call the real and the true principles of knowledge whether of self, the world or the other, real and possible, the unity of theory and practice.

Finally, our human science and civilization had in fact a variety of forms, including the non-dualist, in the pre-modern or the non-modern world, east and west, left and right or classical and vernacular, traditional, official or radical, which the post-modern era should happily restore in fact and value. The recovery of non-western political independence, our *swaraj* or self-rule and self-reform in India (1947) and China (1949) and elsewhere, will then and only then come to signify also the end of western exploitation, domination and hegemony in the sphere of thought, knowledge and culture as much as in the sphere of power.

III

THE TECHNOLOGY OF OBSOLESCENCE IN WAR AND PEACE

It was the British playwright George Bernard Shaw (d. 1950), if memory serves right, who had opined that there is nothing in man's industrial machinery except his (producer's) greed and (consumer's) sloth: 'his heart is in his weapons'. The link between physics (or science) and politics is technology, as Max Born explained in 1960, and it is quite possible that technology, and specially its leading edge, is driven not by considerations of applied science or of commerce but by the allure of armaments that, as we have seen, counts no costs, human or economic. Possibly this is the best definition of the so-called 'dual use' technology, military and civil, which is today leading the world in the fields of nuclear energy, micro-electronics and computers, aerospace, composite materials and biotechnology or specially genetic engineering. At the end of the age of European modernity the structure of its categories, relations and attitudes in both war and peace as well as the dialectics of its changing gear from the one to the other mode of existence can be discerned perhaps together during the short period of the Cold War, 1945 to 1990, and this structure and dialectics we must now review for the sake of prospect as well as retrospect. It is as if the interrelations of science, technology and society are focussed in time of war upon the attitude of expediency, i.e. the end justifies the means as necessary if undesirable, and in times of peace on the attitude of technicism, i.e. the means justifies the end as desirable though not necessary; but in either event there is no principle of unity discernible in this duality of ends and means, desire and power, freedom and responsibility.

The human species appears to be still permanently subdivided in its social organization along two separate and different axes, and according to two independent principles. Firstly, the modern world is divided by the *principle of segmentation* into very many sovereign nation-states or would-be nation-states, large and small, peaceful and warlike, new and old, which are oftentimes grouped into a system of regions, alliances or blocs, west and east or north and south, but all of them represented in individual and formal equality through the institution of the United Nations. Second, the world is divided by the *principle of stratification* into a hierarchy of three classes made up of the industrially advanced market economies or countries, the former centrally planned economies or countries and the backward, underdeveloped economies or the developing countries. The first class of the world system represented about 62 per cent of world manufacturing output, the second class represented about 31 per cent of world manufacturing output and the third class represented nothing very much, except of course tradition, poverty and two thirds of the world's population. In our day, therefore, the world system of modernity was led, at any rate in the chief modern sphere of military-industrial-scientific innovation and diffusion, by a group of six or seven nation-states, the USA, the USSR, Britain, France, China and West Germany (FRG), with Japan perhaps joining along with the USA since 1970 as being the only or the first two post-industrial societies and economies. As anyone can see, these six or seven nations are the same as the ones that fought the Second World War among themselves and involved the world, the first five nations being its victors, now the nuclear weapon haves, and the last two, i.e. divided Germany and Japan, being those two warriors who lost the war and were forcibly disarmed and occupied but won the peace that followed resoundingly. These six or seven states, civil societies and economies, which I call nations, coexisted within a system of relations of their reciprocal international competition and cooperation, forming a coherent segmentary system of the first class, which thus constitutes the external and the internal framework of politics and culture as well as of science and technology among them, within each of them and as the hegemonic leader or the pace-setter for the rest of the world, being the motive power or the power house of modernity and of its end, although not necessarily therefore also of its future.

The two chief social facts or sociological phenomena of military-industrial progress, innovation and diffusion world-wide were (a) that 75 per cent of all military spending and 97 per cent of all military research and development, with science and industry following or working in tandem, occurred in the leading six nations; and (b) that there were alas no product innovations of any consequence to speak of, either military or industrial and civil, occurring in the low-income countries.

I think that there are two separate ways open to us of estimating the profiles of what happens, its costs and benefits, in the life-cycle of innovation, manufacture, use and obsolescence in the world of production, development and progress which we have signally failed to understand, master or domesticate in India. The first method, which is macro-scale and quantitative, will be to take the annual rate of modernization and replacement of fixed capital as an index of the process of self-renewal in the national economy, its conjoint processes of production and reproduction. Thus we find that for the year 1970, say, the annual capital expenditure or investment in new plant, equipment and machinery varied among the leaders from six per cent (USA) to around 15 per cent (Japan) of the GNP. The second method, which is micro-scale and qualitative, will be for us to take as the unit and focus of study, *the project*, i.e. to follow the particular invention from its initial original conception, research, design and development through to setting up for production, start up, fabrication and assembly, marketing, use and final phasing out of the product; and to study a variety of particular cases of individual product innovation in detail, e.g. the Manhattan project. Either method appears, however, to show upon a closer examination that there is always what is called an expressive aspect or element (social, political or symbolic) as well as an instrumental aspect or element (technical, utility or service) involved in this process at every stage; and that this duality applies equally to each new product, to the family of products and to every whole new technology, system of management, know-how or ways of doing things (organization).

It follows from this observation that the calculation of the economic depreciation rate, for example, also includes both sorts of consideration, the technical as well as the commercial (USA) or the technical as well as the social (USSR), although the two aspects or elements were not always easy to separate and recombine. Without being able to undertake this exercise here just now let me say that what I shall define below as the rate of obsolescence (technology, design or fashion) had in fact overtaken the annual depreciation rate (utility, economy or service) in the field of defence quite early on: in value terms the whole stock was being replaced every five years during the Cold War without any use whatever—except to drag the rival potential adversary down the same spiral road (for the UK, 1949–69, see fig. 12).

A case study of
military expenditures in the UK 1949–1969

	Economic life (years)	Scrap value (per cent)	Annual depreciation rate (per cent)
1. Ships	20	10	11
2. Aircraft	10	5	26
3. Tanks & vehicles	10	10	20

Source: SIPRI 1973 : 30n

FIG. 12: The economic depreciation rates of military assets

It is evident that, in any and every modern system, machines have to be replaced almost continually in the industrial man-machine systems, not only or chiefly because they are physically worn out as assets beyond repair after a period of use (the service life or utility time), but also equally or even more because they and/or the products they turn out are after a variable period of time no longer considered socially, politically, symbolically or commercially acceptable as 'up-to-date', i.e. by arguing in respect of their efficiency, performance, design, safety or smartness and other specifications (the time of obsolescence). 'Mostly', stated the *Great Soviet encyclopaedia* without any apparent irony, the 'obsolescence of fixed capital occurs *independently of conditions* within the enterprise employing it' (Moscow,1974, 3rd ed., vol. 16, art. obsolescence; emphasis added). As the necessary if unforeseen consequence of all this kind of thing, which applies just as much or even more in western economic science and technology, I must emphasize that the time of obsolescence, whatever it is called and howsoever it is calculated, has after 1945 for the first time in the evolution of mankind and its material culture, including both its long prehistory and short history, now become remarkably *shorter than the life-span of one generation* of the industrial product, asset or its useful service life, at any rate in all the half-a-dozen leading so-called 'science-based' industries, i.e. nuclear energy, micro-electronics and computers, aerospace, composite materials, biotechnology and of course defence in general, and not only in the vagaries of taste of our perennial women's fashions, the so-called 'rag' trade. This is the most comprehensive and the strictest definition, quantitative as well as qualitative, which

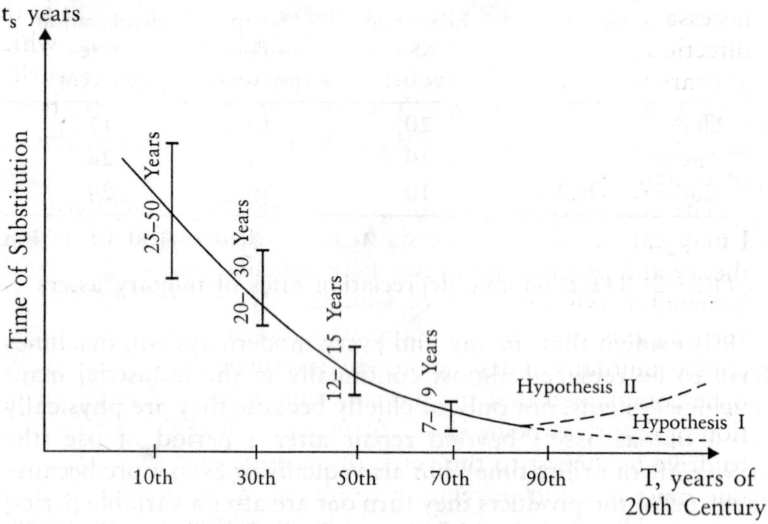

Trends and options in the time of technology substitution

FIG. 13: The leading edge of technology (1979)

Source: Dobrov 1979: 259

I can give of the role, scope and degree of waste achieved in the name of up-to-date-ness, the other side of obsolescence, in the modern way of life and thought.

As it was calculated from the former east, the USSR, the iron law of progress which rules the world today is that contemporary technological innovation replaces and displaces old with new technology, and not just new application or design, after an ever-decreasing interval of time which decays exponentially, halving every 20 years, and which is now seven years, at any rate for its leading edge (G. M. Dobrov, Institute of Cybernetics, Kiev, 1979, fig. 13). This is perfectly consistent in its magnitude and direction of acceleration with the estimate made in the United States at what we may take as the turning point of modernity into post-modernity in 1945 by W. F. Ogburn, sociologist of innovation, who stated then that hitherto it had taken on the average about 35 years for an invention to materialize, and sometimes as many as two or three hundred years or even longer in the past. It still remains for the third world, however, to ask now (a) if this fashionable steep acceleration of the treadmill of progress is altogether benign, necessary and desirable in its scope and effect, (b) what is the direction and meaning of its occult and hidden coupling, which appears to be under no one's control, with men's secret warlike joys, and (c) how soon might it arrive at its inherent limits and the age of technology and the military turn into, say, the age of design and commerce?

The sociology of time in this world system of modernity, if I may call it that, also needs to be worked out in its implicit theory and practice along two perpendicular axes or planes: the horizontal axis or plane of simultaneity (synchrony) and the vertical axis or plane of succession (diachrony), which are of course mutually independent—or interdependent! Goethe had once complained that the human attempt to imagine an operation of nature as being simultaneous as well as successive seems to drive us almost to the verge of insanity, but I think that still one must try, and in relation to human as well as non-human nature. The first plane then is put in place according to the segmentary principle by the mutual relation(s) of two or more players in their common game of whatever it might be in both war and peace. It represents what might be termed the consensus of adversaries or the agreement of competitors, the concept of fashion, trend-setting or plain sibling rivalry leading inevitably in war, or rather more accurately in the preparations for it, to the sequence of new military weapon followed by the counter-weapon and that by the counter-counter-weapon and so on indefinitely. This is precisely the kind of relation that occurred during the Cold War between the two superpowers, the USA and the USSR, and which had led in this period simultaneously to the introduction of much socialist-style bureaucratic planning into the military sector and beyond it in the USA as well as contrariwise to the introduction of principles of capitalist-style consumer or customer sovereignty and even the element of competition into the military sector of the USSR. It has also resulted, for instance, in the introduction and proliferation of all kinds of missiles after the mid-1960s as constituting the fourth arm of the military on top of the army, the navy and the air force in both superpowers, so that it mattered very little incidentally which one of them India decided to follow in her defence plans.

Of course, under the dualistic modern regime of separateness and difference, this apparent convergence of underlying structures in the world system in no way rejoined the cult of the individual in the west and the cult of the collective in the east. Instead, each of them continued to find the other unspeakable as well as insufferable, and the 'iron curtain' was let down across central Europe in order to keep the two of them forever apart in coexistence.

The other half of the self-same world system is set up by the principle of stratification which establishes the perpendicular axis or plane of diachrony and assures the rectilinear conception of modern time versus the cyclical, alternating or rhythmic qualities of time. It owes much of its force to the admirable impulse of upward social mobility, the motive of 'improving oneself' in the world, but now in a new and different sense. It is constructed in the European modernity by what might be called the process of progressive or creative self-destruction, if I may vary a phrase of Schumpeter the economist of innovation, and this aspect of the rat race is now directed against oneself and no longer against any real or imagined other. It is true that in the Hermetic cosmology of Europe, e.g. in alchemy, there is or was said to be cultivated an occult self-rotatory movement in which the subject of regeneration consumed itself, a process often symbolized by the figure of the snake which swallows its own tail. But that cyclical non-dual process was seen and developed within the Hermetic tradition of a certain temporal sequence of the primal unity succeeded by the duality of existence in the world to be overcome by an initiation leading to the subject's symbolical death, rebirth and the experience of restored unity. Schumpeter's explanation in modern economic science of capitalist technological 'creative' destruction is, on the other hand, strictly unilinear and its time flies like an arrow, amounting in effect to technological and organizational innovation for constantly promoting obsolesence of the self as much as of the world of yesteryear, i.e. the formula of self-destruction for the sake of progress and fashion.

The structure and dialectics of the post-war world and of its arms race, dominated by technology, organization and the military, where the emphasis is on accelerated technological progress, are thus to be explained, not by simple adversarial competition alone, but equally by the fact essentially recognized by a group of consultant experts that 'each country is actively seeking the means of defeating its own most advanced weapons and of neutralizing its own most recent defences' (UN centre for disarmament, 1978). It is by the combination of these two complementary aspects of the meaning of time and the effects of technology in the world system that the Soviet Union had come quite close to achieving its cherished Leninist goal of 'catching up with and overtaking the advanced capitalist powers' in the military field at any rate although this process could not produce national security or peace of mind.

Security was in fact unobtainable for either superpower because the arms race after 1945 had already quickly resulted in monstrously huge weapons stockpiles of far more destructive power than had any conceivable military use, the politics of peace apart. The race in the world against oneself as well as the other progressively continued to become ever more complex in technology and organization, firmly entrenched and dangerous as well as costly in military-industrial fashions. This is true of the science and industries of both war and peace, and also of the liminal periods of alternation between the two. 'They do not wait for a potential enemy to react', complained the advocates of disarmament, 'before they react against their own creations.' Thus the per-unit cost of fighter aircraft, for example, was doubling every four or five years, even after inflation correction.

The arms race is by no means over with the Cold War and it has not retracted its attitudes, relations and categories since 1945 but has spread them into peace. The development of armaments is therefore the most potent single evil, peril and bad example that faces the world. It is far more sinful, criminal and dangerous than poverty or disease, the population explosion or environmental pollution, but the men and institutions of science as well as of politics, the expert as well as the elite, still proudly and happily continue to conjointly support and lead it. As I have already quoted for the year 1975 a total of some 400,000 scientists, technicians and engineers, estimated to constitute between one third and one half of the world's trained science manpower, were quite willingly committed to and engaged in new and destructive weapons research and development. In 1983 figures supplied by the US department of defence and the US national science foundation supported the view that the proportion was certainly 30 per cent or higher, and this impression is strengthened in the light of other similar data (OECD). The corresponding labour required for this futility is so highly skilled that it was not unknown in the USA to have to employ one engineer or scientist for every single production worker, e.g. in the Rockwell international B-1 bomber plant at the turn of the last decade (c. 1980), although the prize stealth bomber, claimed to be invisible to radar, is only now unveiled, 1996.

Similarly, in terms of magnitude it has been estimated that of the total cumulative scientific research and development spending in the world since the Second World War altogether some 40 per cent has been directed at achieving military ends in this period of peace. This is the fruit of riches at present and in retrospect under the regime of modernity and post-modernity, east and west, and the world system's middle- and low-income countries had little or no part in it, except as suppliers of its raw materials and some cheap skilled labour in a section of manufactures and services or as customers of its products and designs, mostly the outdated or discarded and obsolete ones.

The latter classes of the non-elite and the non-expert, overpopulated and unenlightened, will always suffer under the rule of the military, technology and fashion as against the rule of design for service or utility, meaning and exchange, but who will bell the cat?

IV

CONCLUSION: THE WORLD SYSTEM

Tiger! tiger! burning bright
In the forests of the night,
What immortal hand or eye
Could frame thy fearful symmetry?

William Blake, 1757–1827

I think that this way of life, mode of thought and code of conduct are far from being over with the Cold War because they had got firmly encoded since the beginning of the Enlightenment in the structure of time itself in the age of European modernity or Kaliyuga, which like a plague of AIDS has gripped all of us from within as well as without. Its first horizontal plane, as we have seen, corresponds to the segmentary principle, the ambition of playing well in the game of modernity internationally, yet also within a framework of national independence, freedom and equality or economic autonomy and political sovereignty. Its second vertical plane corresponds to the ambitions of the stratification principle, affirming hierarchy and snobbery, i.e. keeping up with the Joneses at home and seeking upward social mobility through the path of self-destruction wrongly imagined as periodic self-renewal in the world. In the language of the past Christian life and thought with which it all began in AD 1500, one says that it is as if this modern western civilization had in times of peace as well as war retained its inherited sense of Genesis (time the creator) and of the Apocalypse (time the destroyer), the twofold primordial and dreadful symmetry, but had completely forgotten and lost its balanced cosmological sense of the Eucharist or of the non-dualist mediator that or who sustains and upholds self, the world and the other in the present moment (time the preserver). I ask you now to consider perhaps how much longer such a dualistic system of power and desire, means and ends, creation and destruction, fashion and utility, can possibly continue to sustain—or alternatively to destroy—itself and us.

The telecommunications equipment to be newly installed in New Delhi (1998) according to the consumers' association estimate will have a useful service life of 20 years, and no one has questioned it. According to the practice of MTNL, however, the telephone company would like to recover its capital expenditure money within ten years, say, calculating the payment of the investment or the depreciation at current rates of interest on borrowing. Finally, the technology information, forecasting and assessment council (TIFAC) secretary has stated that the installed equipment will be out of date within five years. In other words, technology, in the sense of engineering, says that the machines are good for 20 years, and economics as the science of scarcity says that they should be paid up with interest in ten years and then scrapped, but the sociology of fashion will declare them to be out of date and wish them replaced within five years. What is the rational method then of fixing the tariff (telecommunications regulatory authority of India, TRAI), and incidentally of proving or disproving the claim of the Enlightenment and of A. P. J. Abdul Kalam that science and technology together are the engine of progress in the national economy as much as in defence?

Personally I foresee in prospect that the *technology of obsolescence* will continue to prevail for some more time, but that it cannot spread much further. In fact it might reach a plateau in the progress of its leading edge fairly soon, but its spirit of militancy, expediency, technicism, obsolescence and waste will surely take over ever new fields of enterprise, physical, chemical, biogenetic and medical, not to mention also the psychological and sociobiological, for any that it might vacate, e.g. the field of passenger aircraft after the Anglo-French Concorde. The rest of the earth will soon find the 'age of technology' quietly displaced and replaced by the 'age of design', which is evidently seller-friendly as well as user-friendly, as for example Agatha Crumm discovers in fig. 14. The businesswoman is of course looking at the manifest face of design and commerce as against the occulted face of the military and hard-science based

FIG. 14: The technology of obsolescence (1989, 1997)

Source: King Features Syndicate, 1989/United Features Syndicate, 1997

THE BORN LOSER by Art and Chip Sansom

FIG. 15: The born loser

Source: Art/Chip, NEA Inc., 1998

technology, but it is for us to ask whether the two can forever march together in dualistic metonymy without exchanging a nod of mutual recognition.

I am also persuaded that the Japanese, who tried very hard, not to speak of us lesser nations, have now abandoned their ambition to lead the development of the fifth generation electronic digital computer in the world. That is, I suppose that the first generation computer was dependent upon the 'cat's whisker' or crystal rectifier after c. 1900, the next similarly on the vacuum tube or radio valve (diode) in the 1920s and 1930s, thereafter developing more rapidly with the now familiar transistor, 1948, the microchip printed circuit, 1960, and successive new generations of silicon chips, all of them western inventions now as before. I cannot say if Japan is secretly planning to elevate the principle of superior design to the status of a new factor of production in order to compensate for its evident lack in the fields of pure invention as well as philosophy. If it turns out to be so, then that would make the factors of production in future eight as against the traditional three: land, labour, capital (pre-modern), enterprise, science and technology (modern), design and finance (post-modern).

The rear end of the world in the future will, therefore, most likely remain to be brought up by that two thirds of humanity, the non-elite and the non-expert, whose millennial hopes will continue to lie in the second coming of Christ or in the next world, unless they should first develop a new science of nature of their own as well as a technology designed for use or service rather than fashion. The leading edge of the elite and the expert will continue its impossible pursuit of power *and* profit, commerce *and* warfare, on behalf of mankind, while secretly yearning for the new Adam and Eve and paradise. Therefore, whether the change of gear marked in history by the fall of the Berlin wall and the end of the bipolar world after 1990 will signify a return to the self-same dualist European modernity of the Enlightenment and its sense of time in a new universal frame or turn the world anew into a self-renewable post-modernity marked by the principles of pluralism and complementarity, combining difference with equality within civil society and among nations or peoples, traditions and states, is surely the open question for the world today. To a possible non-dualist solution of this alternative in relation to self, the world and the other, the subjects as well as the objects of history, we will devote the concluding chapter, but meanwhile it seems that the non-elite and the non-expert two thirds of humanity will have to learn to live, in relation to science, technology and the economy, like the born loser (fig. 15).

5

Self and Other: Dualism and Non-dualism

He has chosen the best possible plan, in which there is the greatest variety along with the greatest order [unity];... the most power, knowledge, happiness and goodness in created things that the universe allowed.

Leibniz, *Principles of nature and of grace*, 1714

The most usual way of representing this unity in duality, [e.g.] linking two groups into one society, is by pairs of opposites... as in the philosophy of Heraclitus [or in the Chinese philosophy]...and the dictum is that *yang* and *yin* together are required to make a unity or harmony (*tao*) as in the union of husband and wife, or the union of winter and summer to make a year....The conception of unity in duality has been used by man not only in the establishment of systems of cosmology but also in organising social structures.

Radcliffe-Brown 1952: 114

What I call the system, truth and method of non-dualism or its corresponding attitudes, relations and categories may be found in a variety of forms, strong like the dialectical synthesis or integration of terms, moderate like correlation, opposition and mediation, or weak as in mathematical set theory. All of them

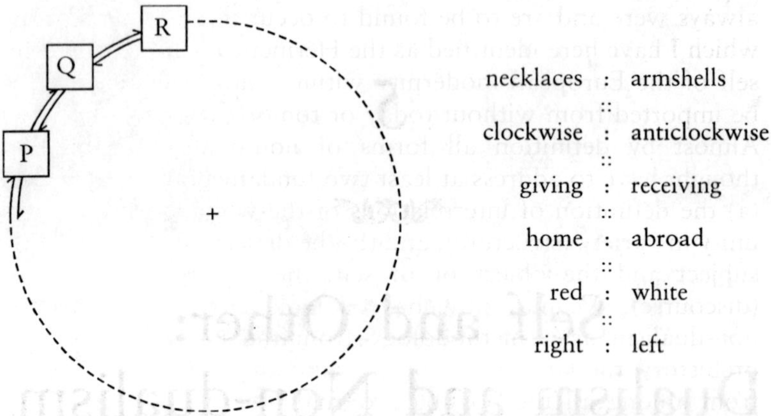

a) Self and other mediated by the gift

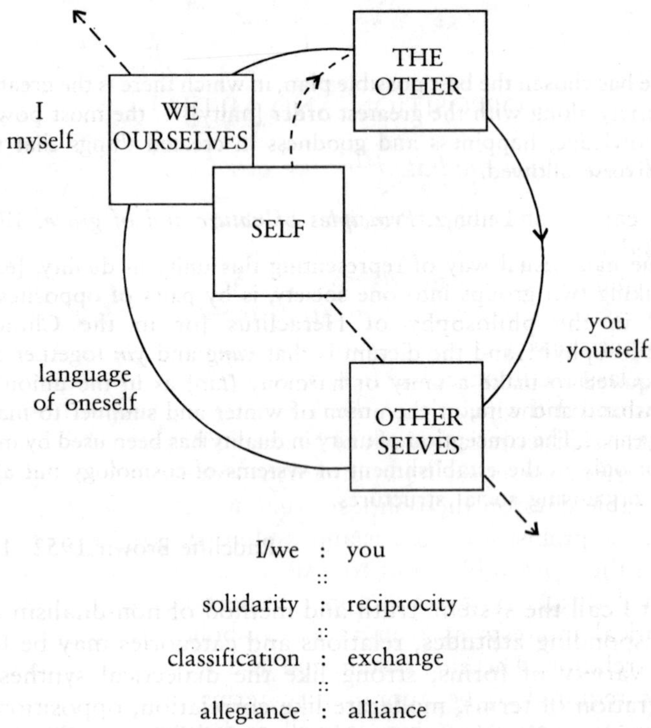

b) Self and other mediated by oneself

FIG. 16: Self and other in exchange theory

always were and are to be found to occur in the other Europe, which I have here identified as the Hermetic tradition, the other self of the European modernity within, and so do not need to be imported from without today or tomorrow or the day after. Almost by definition all forms of non-dualism in life and thought have to address at least two fundamental requirements: (a) the definition of interrelations of the wholes and the parts, unity in variety (structure); and (b) the dialectical relation of the subject and the object or of self, the world and the other (discourse). We may now have a look at three examples of non-dualism as a methodology from the primitive world of prehistory, the language(s) of the child and India, selected here from among all the possible external others of the European modernity.

I

OPPOSITION AND MEDIATION

Self and other mediated by the gift

My earlier 'shining ethnographic detective' work (as Schneider called it in the *American Sociological Review*, 1963, 28: 50) was published as a monograph on the *kula* of Melanesia, and expressed my first attempt to study the interrelations of peoples in politics as well as culture (1962). It was based partly on the classic ethnographic fieldwork writings of Malinowski in London on the people of the Trobriand islands, and partly on the classic theoretical work of Mauss in Paris on the gift in early and tribal societies, its comparative forms and functions. All available data on the politics, ritual and utilitarian trade of the *kula*, a prehistoric inter-island system of symbolic exchange, were there assembled and considered afresh within the contexts of tribal society, giving proper importance to economic and political interests as well as to symbolic values and to their interrelation. The *kula* system and its ceremonial exchange cycle thus required to be appraised in terms of the territorial and political relations within as well as between what are called small-scale stateless societies, applying to Oceania the sociological theory of segmentary systems first developed in relation to

Africa. It was well received as a contribution to the comparative and theoretical study of symbolic exchange and political anthropology, attempting to combine the consensus, conflict and exchange approaches. It taught me incidentally that the ethnocentric hypothesis of human nature is much overdone nowadays, and that the human species is better represented as being in perpetual conflict between its altruistic and its egoistic impulses or principles.

The *kula* of Melanesia is essentially an institution of lifelong alliance and reciprocity set up between two men (it is a men's affair only) who normally owe allegiance and solidarity to two separate distinct societies with somewhat different cultures and languages on two neighbouring islands fixed in a series in space that forms a great closed ring in the South Seas. There are also correspondingly two separate and different kinds of valued symbolic objects, the red shell necklaces and white armshells, which are formally and ceremonially given and received in a kind of dual relay race taking place along each separate chain, single track or continuous ring of individual *kula* partners. The white armshells in effect travel anticlockwise and the red necklaces in effect travel clockwise, and they are kept ideally in perpetual circulation around the ring of islands through the one being always exchanged for the other, but within a system of delayed returns, gift and counter-gift, and never directly. The men of village or island Q, let us say, customarily make two overseas expeditions annually, one to village or island R to go and receive the white armshells and the other to village or island P to go and receive the red necklaces, always setting out from home in their sailing canoes empty handed when facing in either direction, journeying to the left or to the right if one were looking toward the imaginary centre. Similarly, when the men of village or island R come to village or island Q in their turn next year, the men of Q will reciprocally gift them only the necklaces; and when the men of village or island P come to Q separately at their regular appointed time, the men of Q will correspondingly gift them only armshells (see fig. 16a).

As a result, in the long run of the *kula*, all the *giving* of both the white armshells and the red necklaces will always take place at home and all the *receiving* of both kinds of valuables will always take place abroad, the true reciprocity of hospitality, and this rule will hold good for everyone at all times. The two traditionally paired conceptions of 'home' and 'abroad', which are normally mutually exclusive and diametrically opposed for the Pacific islanders who were presumed to be ethnocentrists and cannibals, are in effect brought by the *kula* system of exchange, its symbolism and ceremonial language into a strictly reciprocal and complementary relation in space, time and action, just like the paired conceptions of gift-giving and gift-receiving or of the left and the right, red and white, self and the other, male and female, which are but two opposed aspects of a single unity in duality.

Thus in this kind of prehistoric systems of exchange, whose origins are lost in time, the presumed normal solidarity of the tribesmen of village or island Q and of each island gives way to or alternates regularly in time with a situation in which they all compete politically among themselves within the prescribed framework for the favours of their otherwise alien opposite numbers, the tribesmen of villages and islands P and R. Moreover, by contrast with this lifelong formal ritual alliance or partnership of mutual trust, hospitality and long-term reciprocity, there also takes place customarily under the peaceful umbrella of the same *kula* system and on the same expeditions abroad, the secular exchange of utilitarian objects in something more like an open bazaar or market (*gimwali*). The latter activity is marked by other rules of custom that require each single transaction of profit or loss to be completed almost immediately after haggling and matching in a kind of barter; and the latter trading is kept apart from as well as complementary to the former symbolic exchange by the rule that visitors may only *gimwali* with those natives who are not their respective *kula* partners, allies and hosts. This latter activity then amounts at the most in politics and society to the familiar principle of cooperation among ourselves, solidarity as a group, in order to better compete against others who are strangers, each of the two parties or groups themselves remaining always essentially the same in self-identity, separate and different, at the end of the transaction as they were before, except that their recurrent utilitarian wants of the self, individual and collective, will have been met in the meanwhile.

I should say that the complementary deployment of these two social principles in this prehistoric example of Melanesia (and it could be multiplied from other culture areas) provides us with a clue or a key to the mechanism or one of the mechanisms by which humanity at all times and in all places separates, differentiates and integrates the principles of solidarity and reciprocity, utility and symbol, market and non-market, interest or price and value, home and abroad, gift-exchange and barter or trade. Moreover, in and through this recurrent process or cycle of faith, knowledge and action society and every segment of it continually both forms and reforms itself and also discloses its self-identity to others who do the same. The unity in duality of the *kula* system, far from attempting like money in the theory of the sociologist Simmel to find and establish a general rational medium or the common denominator for all varied individual transactions by intervening at every point between the seller and the buyer of commodities on the market, serves only as a kind of language of mediation, discipline and opportunity to affirm the common humanity of each in indebtedness to all. The message of the *kula* of Melanesia is that, in all human life, society and ceremony, whether it be the aspect of symbol or utility, one exchanges with the other, not only or merely things and signs of the world, messages or values, but also oneself in status, role and self-identity, which are thereby renewed or changed in the process. Indeed, a truly human exchange is possible only through a reciprocity of perspectives between what are otherwise classified as incommensurables, self and the other, individual and collective, the whole and the parts, and not through the imagined equivalence of, for example, gift and counter-gift.

Self and other mediated by the language of oneself

The French psychoanalyst Jacques Lacan, in his recent new reading of Freud, has taught us to regard the unconscious as the discourse of the Other, spelt by Lacan with a capital O, but perhaps it is not the absolute Other that he had in mind with whom it would be impossible to exchange places mutually even in theory. I would rather suppose that, from the point of view of a non-dualist or dialectical semiology which has been my standpoint, the opposition of self and other is mediated by the emergence of the other self and the common human language of 'oneself'. This human language is the real and the true non-dualist locus of culture, labour and politics, whether the other should be God, non-human nature, the world or other human selves, masculine or feminine, native or foreign. The other dualistic view adopted by the European modernity will always lead, as I have argued in the story of the Manhattan project and after, in practice if not always in theory, only to the science of vivisection whose objects will be known and also destroyed or conquered; and equally to the technology of obsolescence or of expediency and technicism, militancy and waste, i.e. to self-destruction of the subject wrongly imagined as self-renewal in the world, the story of Faust or of Dr Jekyll and Mr Hyde, ending in homicide or suicide or both in sequence of time.

Goethe had said that man knows himself only insofar as he knows the world, becoming aware of it only within himself, and of himself only within it. The laboratory of Goethe's culture of knowledge, i.e. the experiment, is (a) *the human body*, which he regarded as the mediation of mind versus the world, as in his example of the (complementary) after-image of the object seen by the eye immediately after it is shut to the world. I have referred to the anthropology of the *kula* of Melanesia to show that the locus of mediation, negotiation and exchange may be also (b) *the body politic*, so to say, and its correlations and oppositions of the internal versus the external, solidarity and reciprocity, power and culture or the individual and the collective, etc. Thirdly, according to the tradition of the Hermetic cosmology and contra Newton, the world is (c) *the body of God*, and the implications of this one can see in the religion, science and medicine and politics of Paracelsus. The point of Goethe's example is that theories of the self, the world and the other should be constructed, represented and considered each of them in the light of the others and in relation to their respective meanings and effects, and not left to remain all three experimentally separate and different, as will happen in the official dualistic classification of the sciences of nature versus the sciences of culture, the two imagined to be stretched horizontally according the polarity of fact and value and suspended vertically according to the polarity of theory versus practice.

The posited primal and simple unity of self and the other is already lost to us and misunderstood, for example, when one says that the mother gives birth to the child, for it would be equally true to say that the child gives birth to the mother, specially if we adopted the view of a social theory of status and roles and not the view of biology. It is surely possible that, in the logical economy of the Christian theology, Jesus the Christ had existed also before the Virgin Mary: perhaps that was why Dante had begun his address to her in *Paradiso* strikingly as 'daughter of thy son'. I think that it would be altogether more accurate, therefore, to say rather that the mother, the child and the language of their relation, separation or negotiation are born of one and the same complex event, which is a dialectical moment of the unity in the duality in its meaning and effect.

Following this original moment of birth, rebirth or creation, the development of the opposition of self and other, the separation and differentiation of 'you' and 'I', is simultaneously always accompanied by the development of a solidarity of the plural 'we', this dual process splitting and initially proceeding in two apparently opposite directions, which will be reconciled by the emergence of the other self or selves. The syntax of the individuation and differentiation of *you* and *I* is the principle of reciprocity, and the corresponding syntax of the identification and classification of *I* and *we* is the principle of solidarity (see fig. 16b). I infer from this observation that reciprocity and solidarity are the two necessarily paired and irreducible principles of society, as against all the other principles of status, wealth and power, hierarchy and authority, domination and exploitation, which only emerge later in the history of the human species perhaps along with the state and classes, etc. The unity in the duality of 'you' and 'I' lies in their mysterious ability to exchange places with one another and even to exchange their respective selves, one for the other, through the conception of oneself, the use of the impersonal pronoun 'one' in language which is employed in English, for example, to refer to either of the two singular persons, 'you' or 'I' or to both of them simultaneously (synchrony) or in a sequence of contexts (diachrony).

By language here is meant something like language when it is paired with logic or mathematics, such as in the proposition that the twentieth century project of the philosophy of language was intended to show that mathematics is just a kind of logic and that logic is just a kind of language, whether this project was successful or unsuccessful (Whitehead, Russell, Wittgenstein). By attempting to make a set of solidarity, classification and exchange or reciprocity, I mean to claim no more than that what might be termed the principle of territorialism is as important as, if not more important than, the principle of domination or the pecking order of sociobiology in the general study of social organization.

In this modern dualistic age, the focus and concentration of scientific genius are, as it were, alternately single-eyed and appear to eschew binocular vision altogether. The French anthropologist Levi-Strauss, for instance, when he deals with the problem of separateness and exchange in society (*Les structures élémentaires*), simply forgets the questions of classification and difference; and vice versa when he deals so admirably with the problem of classification and difference (*La pensée sauvage*), he simply ignores separateness and distantiation as a result. From a non-dualist point of view, however, I must say that the principles of (a) solidarity and reciprocity or (b) classification and exchange, (c) allegiance and alliance, are all simply paired aspects of one and the same attempted solution for the perennial question of human as well as natural, not to say also divine and sacred, separateness and difference (heterogeneity). The solution posited by the Hermetic tradition is, let us say, that self-identity is never self-explanatory but is rather a free gift of the other; and that the unity of the self, far from subsisting in, of and by itself, is truly first and last reducible to the other of its other(s). At any event, I think that such was the lesson of the symmetry of Dante, science of Goethe, medicine of Paracelsus and anthropology of Antigone, apostle of the self-in-society, which I hope to take up on some other occasion.

Towards the logic of unity in variety

We may recall now from Goethe's comparative morphology that the two complementary developments of holism and its apparent opposite, individuation, together form a single dialectical relation, moment or cycle of development in which (a) the successive processes of thesis, antithesis and synthesis, as they occur in the object, are to be followed by (b) the *moment of reflection* in, with or rather through the subject, then followed by the process of elaboration and finally of reproduction and/ or completion of the cycle.

Such a dialectical logic and developmental cycle of unity in variety can be imagined as arising from or itself constituting a four-valued field or system of possibilities, truths if not nowadays also realities. These four are divided and arranged along two intersecting or perpendicular axes, the one of similarity versus difference (metaphor) and the other of proximity or

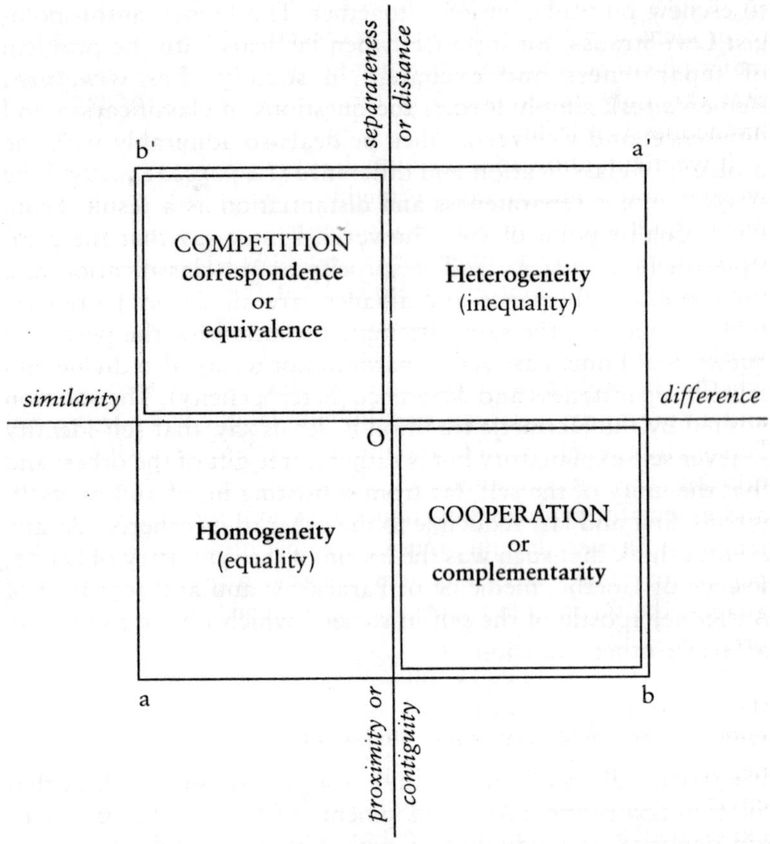

FIG. 17: Structure of national pluralism

contiguity versus separateness or distantiation (metonymy). Modernist thought, life and institutions rule themselves only by the one resultant diagonal opposition of homogeneity versus heterogeneity, respectively identified rightly or wrongly with equality and inequality, whether in one or the other department or entity or between them in severalty. The other diagonal opposition or pair of possibilities would be that of the distribution of any two entities in a relation of their mutual competition, equivalence or correspondence versus the relation of their mutually active cooperation and complementarity. This latter pair of possibilities forming the second diagonal on fig. 17 is almost completely absent or ignored and lies in the shade in our modernist age, although no one has so far cared to produce any logical/theoretical objection to it and there are many phenomenological examples from varied empirical fields that can only be explained in terms of it, e.g. symmetry and handedness (chirality).

If we might allow that the human species and individual men, women and children are or might be *naturally* bilingual or bicultural, then we should see at once that the two states, phases or conditions of homogeneity (equality) and heterogeneity (inequality) cannot properly either limit or exhaust the possibilities of their relations of existence whether in the external world or within the subject, individual and collective, and whether between the subject and the object or, perhaps equally and importantly, between one human language and another. Personally I would include and number under such bilingualism, not only the natural languages of the United Nations, but also and especially mathematics, philosophy and logic, the ordinary language of everyday life, music, dreams, film and the newspaper comic strip, and of course classical and vernacular languages, the written and the spoken, and even silence as well as utterance.

If one is so inclined, I think that one can readily see from the multilingual situation of the city of Delhi, the old as well as the new, that even very young children can happily negotiate this whole field of possibilities in the semiological environment, and even acquire sufficient control to make experiments with it, although this unlisted human ability will be frowned upon officially in school and thereafter, where monolingual standardization is apparently preferred. There is also the historical Indian situation to consider where there was and is, for example, a single system of syntax (Hindustani) with two alternative lexicons, classical allegiances and scripts (Hindi/Urdu), thus making for unity in variety, e.g. using Hindi for prose and Urdu for poetry and song in one and the same Bombay film without complaint. This is as against its possible mirror inverse, which was found in medieval times in another Asian country, where there was almost one lexicon only (Arabic/Persian) but two very different systems of syntax, the Semitic and the Indo-European, an historical situation which certainly allowed a relation of domination, superordination and subordination, but at the expense of unity and the semiotics of intelligibility between the two sides.

The dialogue of *Timaeus* is certainly accepted as Plato's most uninterruptedly known work, which was not lost in the middle ages, and it is therefore to be reckoned as the longest continuously read book of Europe for two-and-a-half millennia. It includes an original theory of vision, which might have been the source of Goethe's secret inspiration, and it is devoted to natural, human and cosmological exploration. The dialogue of *Timaeus* begins with the utterance of Socrates, 'One, two, three—but where, my dear Timaeus, is the fourth...?' I would happily end this defence of what I have called the other Europe and its arts and sciences for the future with a general plea for the four-valued logic of truth and method in place of the restricted two-valued system of dualism that we have inherited from the European modernity. But first I must refer to a long-standing *political* debate relating to the unity and variety of India to which this methodology is pertinent and which might in turn throw some light on it.

II

PLURALISM, CIVIL SOCIETY AND THE STATE

The struggle to define and establish civil society in India during the modern period runs parallel to the rise, development and recognition of the vernaculars and vernacularism everywhere in language, labour and culture; and it is the story of religion and politics proceeding from Kabir (1440–1518) to the martyrdom of Mahatma Gandhi. Its recent political culmination, if we may call it that, is the movement for the linguistic reorganization of the states of the Union after 1950, i.e. as against the previous politico-administrative map of British Indian provinces and the states of the princes. Civil society was thus established in the Indian modernity by the 'separation of powers' between the priest and the prince, the Brahmin and the Rajput or the Thakur, followed by the introduction of a system of pluralism and what might be termed vernacular democracy in the social space so opened for the nation and the people. In this way the path of personal and national salvation in history was simultaneously transformed from *dharmayuddha*, the classical just war of princes and priests, into the modern *satyagraha* defined as the struggle for truth in any and every field by the methods of martyrdom, self-control or self-government and non-violence, whether it was for the achievement of Gandhian independence (*swaraj*), ending foreign domination and exploitation, or for the civic regulation of relations among the different communities at home.

The Indianness of India, under such a regime of pluralism and from a national, structural or relational view, will be found to be inversely proportional to its Hinduness or Hindutva, for example, as this is understood by its chief exponent in the twentieth century. V. D. Savarkar, who hated Gandhi from the time that they knew one another as young Indians in London, 1909, wrote that the strength, cohesion and progress of India depended in the last resort upon the strength of Hindutva, relegating questions of social reform and Hindu–Muslim unity to mere side issues (1923: 119–28). The view that he canvassed at length is entirely European and modernist, not to say Orientalist, in origin and effect. It identifies the strength of any nation or people with the principles of homogeneity or uniformity and centralization, i.e. the coincidence of its land, race, language, culture, religion, etc., culminating in the state. As against this is the pluralistic view that the unity of humanity and of its several segments, secular or religious, depends rather upon the non-coincidence or overlapping of the component races, languages, cultures, classes, etc., which produces or can produce unity through equality *and* difference, relations of mutuality, complementarity and exchange in society, 'unity in variety'.

Even if, by some political process, all non-Hindus were to be extruded from the 'heart of Aryavrata' and its dependencies, they would still remain somewhere next door, and the problem of the relation of self and other, or rather of self and the other self, would still remain to be addressed. This problem of humanity, national and international, cannot be solved within a framework of the majority and the minority, superordination and subordination, the centre and the periphery, the mainstream and the margin or homogeneity (equality) and heterogeneity (inequality). It can be resolved only by the reconciliation or negotiation of equality and difference, the separation of powers, competition as well as cooperation, the convergence of underlying structures or the distribution of differences into complementary domains as the principles and bases of mutuality, reciprocity and exchange, meaningful and effective, whether it be in amity or enmity.

Savarkar, whether or not he was the inspiration of terrorists and assassins, brought all his talent and energy to focus on the project expressed by the slogan that he coined on his birthday in 1941, 'Hinduize all politics and militarize Hindudom!' He was widely regarded as a hero of the national freedom movement; and for him the Hindus by themselves constituted 'the foundation, the bedrock, the reserved forces of the Indian state'. Gandhi, on the other hand, who led the movement in civil society, was the martyr of *swaraj*, self-rule or home rule. Even during his campaigns of civil disobedience he was the supreme witness to the causes of pluralism and social reform, Hindu–Muslim unity, the removal of Untouchability, and of *swadeshi* (home-grown) as the love of one's neighbour and of his or her labour. For all the lip service paid to him after the event and also to his legacy of method, the theory of *satyagraha* or putting the truth of God and society above the state as well as above the tradition of the priestly *shastras*, Gandhi now lies more or less rejected in India. Savarkar saw the strong Hindu state as being effective in offence—rather than only in defence—and capable of dictating its terms to the world regarded as a system of states or of nation-states, with scarcely a thought for society or reform, let alone for pluralism and Hindu–Muslim unity.

Gandhi wanted the pluralist 'equality of all religions' (1930) as the foundation of modern Indian society, nationality and culture, whether politically organized as one independent state or as two (1947); and he consistently refused to subordinate means to ends or ends to means, either for himself and his own or in relation to any other segment of humanity, which he regarded as a system of national and international societies, peoples and cultures or traditions like languages. Needless to say, Savarkar rejected outright the truth, method and praxis of *satyagraha* or Gandhian non-violence, and declared it to be suicidal as well as sinful.

The difference between these two national leaders from the standpoint of the present study lies equally in their corresponding attitudes to European or western modernity. Savarkar's father was a landowner known for his western-style education as well as his Sanskrit scholarship, and the son brought equal disgrace to both traditions. He and his 'new India' group, fascinated by the prospects of politico-military action as the way of national salvation (*dharmayuddha*), went and secretly learned the art of bomb-making from a failed Russian revolutionary in Paris. Indeed, this strain of Hindutva, because it is fearful of dialogue, will always want to combine India and Europe, tradition and modernity, in a non-relation of pure metonymy and perfect dualism, without any mutual conversation and without any sense of reflexivity or self-criticism applied to either point of reference. In other words, it must always implicitly depend, like all projects of political communalism or revivalism—Hindu, Muslim or Sikh—upon an imported modernity, since it is incapable of producing any principle of motion from within itself.

Gandhi, on the contrary, would always look the other in the eye as his second self and offer truthful dialogue and non-violent conversation without fear of the possible consequences. For him the national freedom movement for *swaraj* meant essentially the self-reform and self-rule of civil society; and Ram Rajya was to bring the rule of salvation in society, a 'kingdom of Heaven' in politics viewed as the self-management of the institutions of civil society rather than of the state. While telling the British as well as Tagore that 'our non-cooperation is a retirement within ourselves' (1921), he in fact produced out of tradition or rather from the custom of the country a modernity of India's own and so that India should come into its own in the world with a reformed tradition of pluralism and what I would call vernacular democracy. Incidentally, this was also the moment of perfect collaboration between non-cooperation as the form of the national freedom movement, on the one hand, and the self-reform of Sikhism, 1920–25, on the other, by the methods of *satyagraha*, martyrdom and non-violence.

The same attempted combination of pure anti-Muslim Hindutva, passed off as tradition, and a juxtaposed mindless pro-western modernity already finds expression in the Bengal renaissance of the century before. Its greatest luminary of the late nineteenth century was Bankim Chatterjee, 1838–94, who both presents and explains the mystery in the last chapter of his most important work, the novel *Anandamath*. Here the two intersecting dualisms of fact and value, 'physical' and 'spiritual', and of theory versus practice, 'knowledge' and 'action', which constitute the specific matrix of the European modernity as we have argued in chapter 2, are fully internalized and presumed by Bankim to be self-evident universals of truth and method. And the looked-for saviour is again the state, and not any change of civil society, no matter that the state must be a foreign body, as in speaking here of the *sannyasi* revolt, c. 1770.

> Your task is accomplished. The Muslim power is destroyed.... What is going to happen now is for the best.... True Hinduism consists in knowledge, not in action. Knowledge is of two kinds, physical [fact] and spiritual [value]. Spiritual knowledge is the essential part of Hinduism. If however physical knowledge does not come first, spiritual knowledge will never comprehend the subtle spirit within. Now physical knowledge has long since disappeared from our land, and so true religion has gone too. If you wish to restore true religion, you must first teach this physical knowledge. Such knowledge is unknown in this country because there is no one to teach it. So we must learn it from foreigners. The English are wise in this knowledge and they are good teachers. Therefore we must make the English rule. Once the people of India have acquired knowledge of the physical world from the English, they will be able to comprehend the nature of the spiritual. There will then be no obstacle to the true Faith (*Anandamath* or *The Abbey of Bliss*, 4: 8, trans. T. W. Clark *in* de Bary 1958: 715f).

The life of the sacred, civil or moral society in Islam, I would hasten to add, was equally lost in the period of medievalism somewhere between the priest and the prince, an hypotrophied tradition, on the one hand, and an hypertrophied state, or at any rate its executive branch, on the other hand. For further study one will have to determine the elementary structure or principles of this unfortunate dualist condition, Hindu and Muslim, and then return with more confidence to the problematic of the Indian modernity of salvation in society and history and the non-dualism of unity in variety.

The chief difficulty about introducing Sikhism, for example, at the beginning of the modern period and Gandhism at the end in this universe of discourse as new modern forms of non-dualism, salvation in society and history, is surely one and the same. It is due to an unacknowledged and unexpressed principle of closure, a bar which is widely assumed, whether consciously or unconsciously, by historians as well as theologians. Around AD 1200, if we turn to Europe, it was understood that there was to be no new religion or revelation, such as the second coming or the tenth *avatar*, and any novelty of religion was to be classified, therefore, as being merely schism, heresy, religious protest or reform. The Christian pope, Innocent III, relying upon both imperial and canonical tradition in his decretal, *vergentis in senium*, addressed to the clergy and the people of Viterbo, for the first time identified heresy with treason, i.e. from the view of the established single religion of the priest and the prince between whom lay the potential space of civil society. After 1208, the spiritual sword of dialogue was finally cast aside, and papists and statists were united in the view that 'iron was to conquer those whom persuasion would not convince', thus combining religion and politics in the worst possible way and to the utter detriment of civil society.

In the classical or medieval view that is still held by many, then, the doors of revelation and interpretation were closed by authority long ago in history, and no one can now seek to establish in the world a (new) society for God-realization or self-realization. F. Schuon, a respected and learned student of comparative religion, having confessed his adherence to this standpoint of the closing of the gates (say) after Islam, or in general the middle ages, then vainly tries to make a sole exception in favour of Sikhism as an example of the 'very latest possibility' and that also made possible by the quite exceptional context, as he says, of the proximity of Sufism and Hinduism (1963: 50n). But I hold and would urge to the contrary that the coexistence, dialogue and exchange of aspects among religions are typical of humanity and civil society, and by no means exceptions in history as distinct from historians.

For it is a law of religion and society, and not only for the Christian martyrs, that whosoever tries to save his self shall surely lose it, but the one who will lose himself for the sake of truth, God and other human beings or nature, shall find his true self in love of the other. I cannot imagine why we ever needed the British to divide and weaken our society, while pretending to strengthen the state, when the prevalence of the one true religion, say, of the khaki shorts and the saffron flags or Hindutva, as pitched against the one true religion of some other Indian, usually Muslim, can do it so much better and simultaneously leave the modernity of science, technology, organization and economy dependent upon foreign tutelage, inspiration and control or leadership.

The nature, praise and pursuit of the national civil society in power as well as culture can be put in the form of four propositions. (a) Civil society is truly the locus of God-realization or self-realization as well as of the common usage or custom of the people, and not only of conformity to authority and tradition or what is handed down. (b) It is the sovereign arbiter of custom as against the priest, custodian of tradition, and the prince, maker and executor of the state and its law. (c) Inspired in the modern world by new religion or secularism or, as I think, pluralism, civil society alone has the inherent power to find a people's principle of history and so to change the common usage or 'custom of the country' and itself. That is the modern and Gandhian realization of *swaraj* or self-rule and self-reform, the one being the condition of the other, in civil society, which is established by custom and the common usage in the space opened between tradition and the state or the priest and the prince, and it is surely the good way to combine religion and politics in the management of institutions. The Hindus, whether or not they were given the name by the Muslims and their sacred Vedas recovered for them in the written form by

Europeans who were Orientalists, are no more (or less) qualified in themselves than non-Hindus to speak for India under the regime of pluralism; and whether or not (say) the Sikhs and the tribals are to be classed as Hindus is a purely subordinate question that we shall not attempt to answer. (d) The prince and the priest, whenever the two rule together, either through a state-established religion or a religion-established state, are the enemies of civil society, its national autonomy and self-expression, political economy, customs and morality.

III

UNITY IN VARIETY

The weight of my sociological and ethnological studies has slowly forced upon me the conclusion that the plurality or variety of traditions, a community of sovereignties, is a great human achievement in the evolutionary discourse of history, politics and society; and that the system of national pluralism as the mediation of the one and the many has always formed part of our most precious heritage in India, primitive or prehistoric, ancient, medieval and modern. It is not susceptible of future post-modern recognition and development, however, within the framework of either a system of domination, stratification and hierarchy, or alternatively of a falsely imagined unity achieved through uniformity, homogeneity and centralization. There is indeed a complex problem here of similarity and difference, proximity and distance, or of interaction and innovation, competition and cooperation, in the interrelations of (say) Hinduism, Sikhism and Islam, not to mention the other traditions. This should be seen, however, within the philosophy and the institutions defined in the Indian modernity by Mahatma Gandhi under the heading of the 'equality of all religions', i.e. all of them equally true and truly equal as well as all equally humanly imperfect, and therefore each of them capable of self-reform as well as self-rule (*swaraj*) in the common or transcendent/immanent light of God's truth and method of non-violence, the fatherhood of God and the community of humanity. It is useless to try and distinguish in the study of comparative religion between the so-called revealed religions and the others, whatever they may be named, since God cannot reveal a religion, but only himself, and all religions are therefore in some sense man-made in history and society.

Apart from or along with the hardy perennials of domination and exploitation, death and destitution, affliction and alienation, freedom and responsibility, the problem for the world today is perhaps the reconciliation of difference and equality in civil society. This surely cannot be achieved within a framework of the (permanent) majority and the minority, superordination and subordination or the centre versus the periphery, but will require embodiment of and participation in the logic of pluralism, mediation of the one and the many, which we call in India 'unity in variety'. It is hardly necessary to add that to this Gandhian attitude in religion, for example, there would correspond relations of a true federalism in the state, even if very imperfectly under the Mughals and the British, and of real tolerance and purity in civil society and the national economy, which is the modern Indian way of learning when to combine as well as when to separate a multi-religious nation, a modern pluralist society and a federal secular state.

I should now explain the background of the Indian modernity according to the rules of semiological method, i.e. on the analogy of language, which is the people's system of custom or the common usage that is furthest removed from the influences of the priest and the prince, the authority of tradition as well as of the state. The system of language represents the country, as it were, in the ancient trinity of God, king and country which is familiar from Europe as well as India.

> Linguistic diversity is an established way of life in India;... which is, in fact, one aspect of cultural pluralism. People in a multilingual country develop ways of speaking in which they can maintain their own language and still communicate with speakers of other languages; they become bilingual or multilingual. Languages are not employed parallelly in all spheres of activity,... the patterns of use of languages in a [single] bilingual community are largely complementary.
>
> All languages are equal, but different languages have different roles in society, and have different positions in the education system.... That each speaker has [reciprocal] access to the other's variety (e.g. Hindi or Urdu) and also to the common variety (Hindustani) underscores the integrative aspect of such variations [v. uniformity]. The assertion and acceptance [of identities] are reciprocal and mutual. In this way,

different social groups can maintain their separateness on the one hand and express their togetherness on the other. The face-to-face communication allows for more [strategies of] negotiation than any other context.

...[Pluralist] integration is thus different from [a] 'melting pot' on the one hand and [b] segregation on the other. Melting pot results into complete assimilation with the dominant group, a merger of identity. Segregation results into isolation [or stratification] and the tensions thereof.

Hindi and Urdu, in the language of structural linguistics, form one set of correlation; it is an opposition in which one can only be explained in the context of the other: Hindi and Urdu are two sides of the same coin: if you have one then you have to accept the other.... Urdu identity [for example] becomes more relevant only in the context of Hindi rather than in the context of any other language.

Though the institutional affiliation of Urdu is with the Persian literary tradition and that of Hindi with the Sanskrit literary tradition, Hindi and Urdu literature have been developed by the same population in the same region—Hindu and Muslim writers have contributed to both these varieties. As in various other spheres of creative arts, music, painting, dance and drama, so [also] in writing, there has been a remarkable reciprocity.... Persian scholarship flourished alongside the traditional Sanskrit scholarship for a long time in the Northern courts...[and] for the first time in the history of Indian languages, a common *lingua franca* [Hindustani] becomes the language of the court, administration, law and order though spread [and varied] over a vast territory.... Linguistically, Urdu and Hindi are recent [alternative] standardisations of a medieval [or early modern] *lingua franca*.

Our linguistic issues are not the multiplicity of languages, social variation and varieties, but of the gaps between the colloquial [or the vernacular] and the [elite] literary standards (Pandit 1977: 3f, 37, 47, 50, 56, 60).

The logic of national pluralism, e.g. functional multilingualism or multiculturalism, makes the common usage of civil society prevail over or along with the authority of inherited tradition and/or foreign imports, and perhaps this is the normal and the proper condition of modernity. The pattern of multilingualism is therefore inclined to keep its distribution of differences generally functional, contextual and domain-specific as well as group-specific, e.g. in relation to the family, neighbourhood, bazaar, formal education, the library and the laboratory, work, administration, worship and sport or play, although there will be of course degrees of overlap and competition. Thus several independent empirical studies confirm that written as well as spoken languages live, grow or spread and develop, not by seeking after purity or insulation and officially sponsored standardization, but by ongoing free processes of contact, reciprocity, interaction and exchange with other languages in a plural setting or context of situation, national and international.

The European writers who could be quoted on the subject, e.g. French and Slavonic, have duly recognized but two tendencies at work in history and society, the centripetal and the centrifugal, and those two mutually opposed and unmediated, whereas I think that India will show us a field of four possibilities to make up the logic of pluralism as the mediation of the one and the many, unity in variety.

> Humanity is constantly struggling with two contradictory processes. One of these tends to promote unification, while the other aims at maintaining or re-establishing diversification. The position of each era or of each culture in the system (the orientation in which it finds itself engaged) is such that only one of the two processes seems to have a meaning. The latter is seemingly the negation of the former,...we are dealing with two different manners of *making oneself* (Levi-Strauss 1977: 361).
>
> In the union of different levels and subsystems into a single semiotic whole—'culture'—two mutually opposed mechanisms are at work: (1) the tendency toward diversity [heterogeneity]—toward an increase in differently organized semiotic languages, the 'polyglotism' of culture; (2) the tendency toward uniformity [homogeneity]—the attempt to interpret itself or other cultures as uniform, rigidly organized languages (Lotman *et al.* 1975: § 9).

That is to say, our modern Europeans can only imagine any two entities to be dualistically related either as similar and together, producing homogeneity and equality, or else as separate and different, producing heterogeneity and inequality. But this opposition wholly ignores, obscures and excludes utterance of the other pair of logical and empirical possibilities that the two entities might be in the relation of correspondence, equivalence and competition, i.e. separate but similar, or else of complementarity and cooperation, i.e. different and together.

All variation of custom or the common usage, taking primarily the analogy of language, which gives the alternative appearances of distributions that are either on a vertical plane (stratification) or on an horizontal plane (segmentation), can be resolved and united into a single systematic field of cultural/ political pluralism or 'unity in variety'. The two diagonals shown in fig. 17 thus represent respectively the theory of social stratification or of equality and inequality (aa') versus the theory of a division of labour, as it is called, segmentation and correspondence or complementarity (bb'). We can resolve both of them simultaneously into the two fundamental axes as shown, horizontal and vertical, i.e. respectively of similarity/ difference and of proximity or contiguity/separateness or distance, and so term the whole reconstituted single system or unified field of self-identity, classification and exchange, 'unity in variety'. This is the structure of national and international cultural/political pluralism, as explained further elsewhere, and one of its remarkable effects is mutual convergence as well as divergence, the emergence of varied, but that is not to say disparate, modes of human articulation and communication (Uberoi and Uberoi 1976).

P. B. Pandit's new contribution to sociolinguistic theory lies, then, in his studies of what we have termed the horizontal plane or axis of speech varieties. We have inferred that multilingualism, when taken as a characteristic Indian and human way of life, is constantly produced, sustained and made functional or articulated and integrated by virtue of the simultaneous operation of two general social tendencies or principles. (a) The distribution of differences into complementary, rather than only into competitive, domains of activity or contexts of situation. (b) The convergence of underlying structures, and not only their mutual divergence, in history and society through the free human acts of communication, interaction and exchange.

For example, the Saurashtri dialect of Gujarati emigrants who settled in Tamil Nadu from c. 1500 onwards is Pandit's case in point. The similarities between Tamil and Gujarati can be seen as part of a wider framework of comparison of Dravidian and Indo-Aryan. There are certain significant differences also between the two languages and language groups. Insofar as Saurashtri shares these distinctive features with Tamil and not with Gujarati we are able to say that the linguistic structures of Saurashtri and Tamil are much more closely related to each other—indeed they are almost identical—than either of them is to the original parent Gujarati.

The first social principle, therefore, is a kind of self-control on evolutionary variation and differentiation within a system (synchrony), while the second is a self-control on variation or diversification between two or more systems in evolution (diachrony). We should conclude that, under a regime of pluralist non-dualism, all human differences and partitions are negotiable in civil society as a 'community of sovereignties' because no one reality or truth falsifies another. Thus the non-dualist attitudes, categories and relations or the strategies and the dialectics of negotiation, whether it be in amity or enmity and in harmony or conflict, can together explain the processes in history and society of the convergence of underlying structures and the distribution of differences into complementary domains. In effect our common humanity thereby returns to the perennial fashion of the Hermetic tradition of Europe, and produces neither simple homogeneity (equality) nor heterogeneity (inequality) but a new non-dualist axis of correlation and mediation, correspondence and complementarity. From the standpoint of pluralism, one may say also that the principle of the 'separation of powers', within and without the state, is merely the other half of the principle of the division of labour in the economy, and both are aspects of unity in variety. Finally, the theory of evolution means to us, not chiefly or only the development of what is complex out of what is simple, but also and equally the development of many varieties of existence out of the original few, and without humanity in any way losing the unity of its universe of discourse.

References Cited

BIRNBAUM, NORMAN. 1959. The Zwinglian Reformation in Zurich. *Archives de Sociologie des Religions* (Paris), 8: 15–30.

BOETIUS, HENRIK, MARIE-LOUISE LAURIDSEN and MARIE-LOUISE LEFEVRE. 1998. *Lyset, morket og farverne* (Light, darkness and colours). Copenhagen: Multivers.

CHURCHILL, WINSTON S. 1954. *The Second World War*, vol. 6. London: Cassell and Co.

CLARK, RONALD W. 1971. *Einstein. The life and times*. New York: World Publishing Co.

DANTE ALIGHIERI. 1908. *The divine comedy*, trans. H. F. Cary (Everyman's library). London: J. M. Dent and Sons.

DE BARY, W. T. *et al.*, eds. 1958. *Sources of Indian tradition*. New York: Columbia University Press.

DOBROV, G. M. 1979. A strategy for organized technology. *Technological Forecasting and Social Change* (Elsevier North-Holland), 13: 257–71.

DUBY, GEORGES. 1981. *The age of the cathedrals*. Chicago: University of Chicago Press.

GROUEFF, STEPHANE. 1967. *Manhattan project. The untold story of the making of the atomic bomb*. London: Collins.

HASTINGS, JAMES. ed. 1912. *Encyclopaedia of religion and ethics*, vol. 5. Edinburgh: T. and T. Clark.

HERMETICA. 1985. *Hermetica. The ancient Greek and Latin writings which contain religious or philosophic teachings ascribed to Hermes Trismegistus*, 4 vols, ed. and trans. Walter Scott with addenda by A. S. Ferguson. Boston, Massachusetts: Shambhala. First published by Oxford University Press, 1924ff.

LAQUEUR, THOMAS. 1987. Orgasm, generation and the politics of reproductive biology. *In* Gallagher, Catherine and Thomas Laqueur, eds, *The making of the modern body*, pp. 1–41. Berkeley, California: University of California Press.

LEIBNIZ, G. W. 1900. *The philosophy of Leibniz* by Bertrand Russell, with an appendix of leading passages, pp. 203–305. London: Allen and Unwin.

LEVI-STRAUSS, CLAUDE. 1977. *Structural anthropology*, vol. 2. London: Allen Lane.

LOTMAN, J. M. *et al.* 1975. Theses on the semiotic study of culture. *In* Sebeok, T. A., ed., *The tell-tale sign. A survey of semiotics*, pp. 57–83. Lisse, Netherlands: Peter de Ridder Press.

MORRISON, PHILIP. 1946. The laboratory demobilizes. *Bulletin of the Atomic Scientists* (Chicago, Illinois), 2(9, 10): 5–6.

NEWTON, ISAAC. 1931. *Opticks*, 4th ed., 1730. London: G. Bell and Sons.

PAGEL, WALTER. 1958. *Paracelsus. An introduction to philosophical medicine in the era of the Renaissance.* Basel and New York: S. Karger.

PANDIT, P. B. 1977. *Language in a plural society*. Delhi: Dev Raj Chanana memorial lectures/Delhi University Press.

RADCLIFFE-BROWN, A. R. 1952. *Structure and function in primitive society*. London: Cohen and West.

RUSSELL, BERTRAND. 1949. *The scientific outlook*, 2nd ed. London: Allen and Unwin. First published 1931.

SAVARKAR, V. D. 1923. *Hindutva*. Nagpur: V. V. Kelkar.

SCHUON, FRITHJOF. 1963. *Understanding Islam*. London: Allen and Unwin.

STOCKHOLM INTERNATIONAL PEACE RESEARCH INSTITUTE (SIPRI). 1973. *The meaning and measurement of military expenditure*. Stockholm: Sipri/Almqvist and Wiksell International.

UBEROI, J. P. S. 1962. *Politics of the kula ring*. Manchester: Manchester University Press.

UBEROI, PATRICIA and J. P. S. UBEROI. 1976. Towards a new sociolinguistics. *Economic and Political Weekly* (Bombay), 11(7): 637–43.

VERNANT, J.-P. and P. VIDAL-NAQUET. 1972. *Mythe et tragédie en Grèce ancienne*. Paris: Francois Maspero.

WEYL, HERMANN. 1952. *Symmetry*. Princeton, New Jersey: Princeton University Press.

Index